A Compendium of Deformation-Mechanism Maps for Metals

David J. Fisher

Copyright © 2022 by the authors

Published by **Materials Research Forum LLC**
Millersville, PA 17551, USA

Published as part of the book series
Materials Research Foundations
Volume 116 (2022)
ISSN 2471-8890 (Print)
ISSN 2471-8904 (Online)

Print ISBN 978-1-64490-168-7
ePDF ISBN 978-1-64490-169-4

Distributed worldwide by

Materials Research Forum LLC
105 Springdale Lane
Millersville, PA 17551
USA
http://www.mrforum.com

Printed in the United States of America
10 9 8 7 6 5 4 3 2 1

Table of Contents

Table of Contents

A Compendium of
Deformation-Mechanism Maps for Metals

Introduction

Equilibrium phase diagrams are an invaluable guide to the obtention of metallic structures having a desired structure, even when the preparation conditions are quite far from equilibrium. Having created such structures, deformation-mechanism maps should be an equally invaluable guide to predicting the optimum processing conditions for a material. But following an initial enthusiasm for the plotting of such maps, their use is now surprisingly limited.

There exist voluminous encyclopaedic maps of processing conditions, usually in the form of strain-rate versus temperature. A typical example is the comprehensive volume edited by Prasad et al.[1] Although they are useful, such strain-rate versus temperature plots fail to capture the more fundamental aspects of deformation because there are so many variables which vary from material to material.

Deformation-mechanism maps summarize the fields of stress and temperature within which a particular plastic-flow mechanism predominates. Most materials are able to deform via alternative independent mechanisms, such as dislocation glide, diffusional flow and dislocation creep. Each of them appears as a field on the map. A given point then predicts the dominant mechanism and the associated strain-rate. The maps permit study of the effect of crystal structure and atomic bonding upon plastic flow. They can guide the design of experimental studies of a given flow mechanism and also the identification of missing mechanisms; rather in the manner that the periodic table of the elements predicted missing entries. They are also useful in matching a material to a given engineering application. They are particularly useful when they are related to the most fundamental properties.

It helps, for example, to normalize the temperature by dividing it by the absolute melting-point. There is, after all, the metallurgist's rule-of-thumb that deformation at an homologous temperature which is higher than 0.5 counts as 'hot-working'; even if the actual temperature is the ambient temperature in the case of lead, for example. It is also useful to divide the applied stress by a suitable elastic constant. This again permits a more meaningful comparison to be made between the deformation-mechanism maps of differing materials.

The dominant mechanism which controls deformation changes from point to point over the map, and its predominance depends very much upon stress, temperature, grain size, dislocation density, *etc.* Weertman was the first[2,3] to suggest the construction of such maps, and his original efforts concentrated on creep phenomena, with the normalized stress being plotted against the homologous temperature for a given grain size (figure 1a). Such maps can be constructed by using known constitutive equations to describe a given mechanism. The map is then naturally divided into various fields, depending upon which mechanism governs deformation.

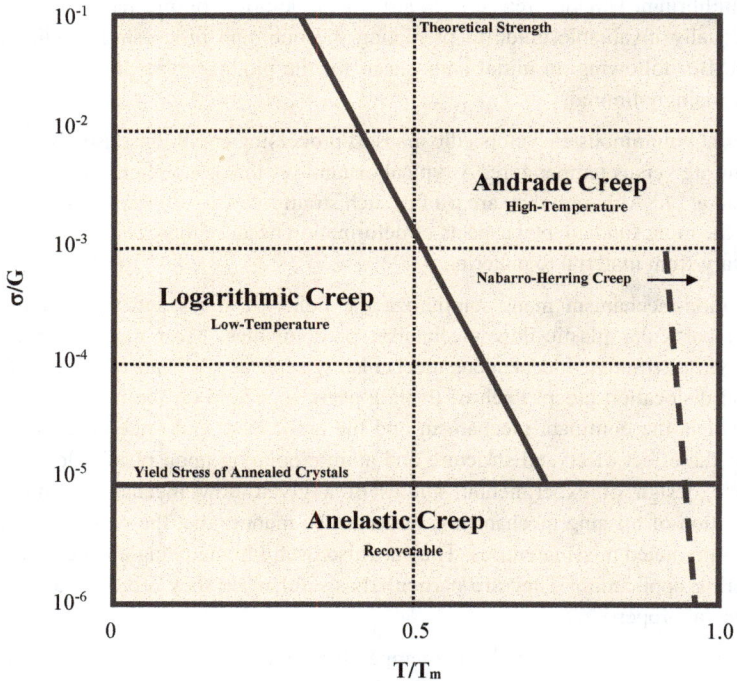

Figure 1a. An early deformation-mechanism map

It was then shown by Ashby that the deformation mechanism map could be immediately simplified by plotting the normalized stress versus the reciprocal of the homologous temperature. This caused the boundaries which separated the various fields to become straight lines. Constant strain-rate contours could also be approximated by straight lines.

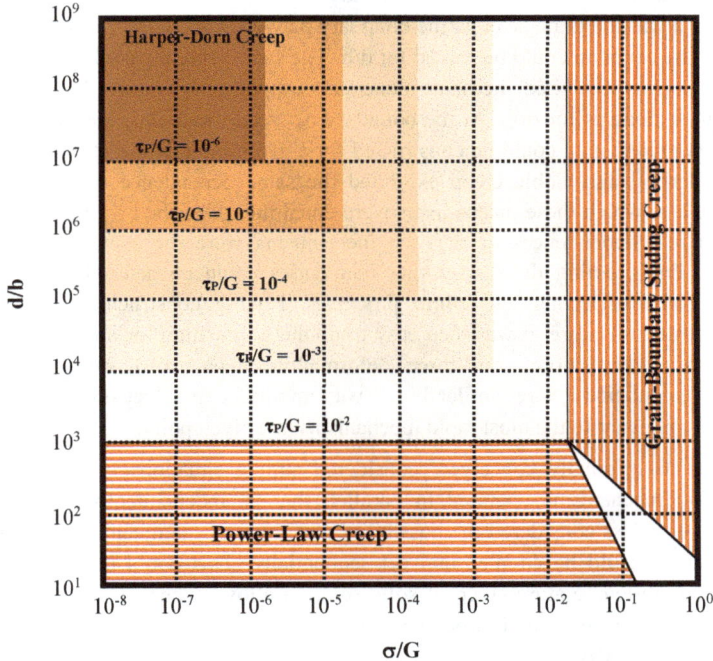

*Figure 1b. Expansion of the Harper-Dorn creep field
with incorporation of the Peierls stress*

A new form of map was later introduced which was based upon the grain size. The Weertman and Ashby maps of normalized stress versus homologous temperature had two problems. One was that they were relevant to a given grain size but covered a huge range of temperatures; even though the choice of temperature was not usually in the gift of the engineer for a given application. The other problem was that the various fields were again separated by curved lines. Determining the position of the latter offered considerable mathematical difficulties.

The construction method for the new maps, taking account of only climb, Nabarro-Herring and Coble processes, involved for example the comparison of two mechanisms which exhibited the same dependence upon stress but which involved differing diffusivities and differing power-law dependences upon b/d, where b was the Burgers vector and d was the grain size. At the boundary between two fields, the strain-rates were equal and the equation could be solved for d/b. The same principle could be applied when two mechanisms exhibited the same dependence upon temperature, but had differing power-law exponents for σ/G. At the boundary between two fields, the strain-rates were equal. A complete map could be constructed for a given high temperature by noting that Nabarro-Herring and Coble creep exhibited the same dependence upon σ/G, and the value of d/b at which these mechanisms were equal then furnished a value of d/b. Given that the stress dependences of the two mechanisms were the same, this provided a horizontal line crossing the map. Other boundaries could be determined by similarly equating Nabarro-Herring and climb processes. Having constructed a map for one homologous temperature, it was then easy to obtain a new map for another homologous temperature. When more than three deformation mechanisms were involved, the necessary calculations were similar but it was important to select only the boundaries which were relevant to the most rapid mechanisms at a given point.

The maps, as originally conceived, considered simple situations in which individual deformation mechanisms operated in parallel; that is, independently of one another. Ashby nevertheless considered the possibility that some of the mechanisms could act in series; that is, could be coupled and act sequentially. This could occur, for example, during the high-temperature creep of solid-solution alloys, where dislocation climb and viscous glide are sequential processes. Its effect can be seen in the deformation-mechanism map for Al-3at%Mg.

A valuable feature of the maps was that their comparison often clearly reveals how the various fields change in shape, size and position as a function of changes in other variables. For example, the Peierls stress is an important factor which controls the transitions between Harper-Dorn creep, grain-boundary sliding and power-law creep regimes. The effect of the magnitude of the Peierls stress upon the boundaries between the various mechanisms can be clearly seen in a deformation-mechanism map (figure 1b), which dramatically reveals that Harper-Dorn creep can be the dominant mechanism over wide ranges of grain size and stress[4] for a given material. Its effect can also be compared for a wide range of materials (figure 1c)[5].

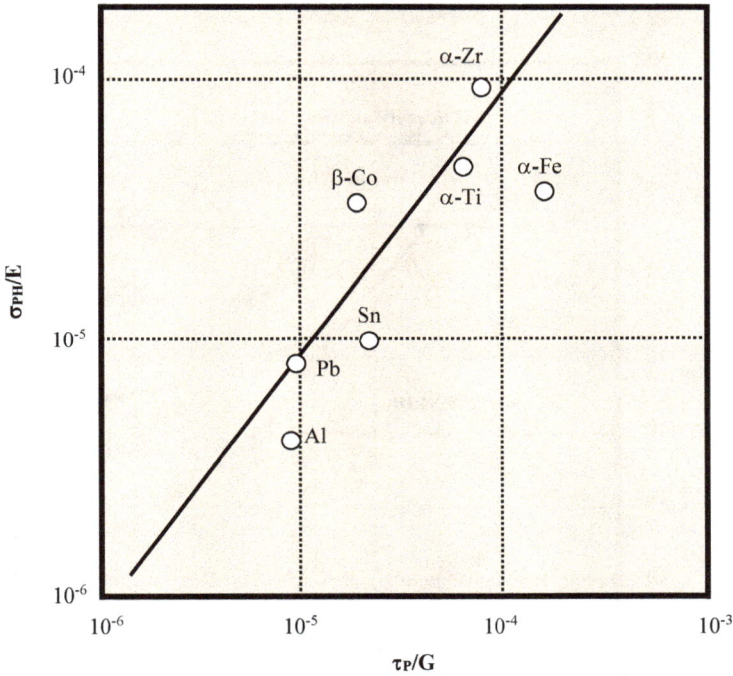

*Figure 1c. Dependence of the transition from power-law
creep to Harper–Dorn creep upon the Peierls stress*

In spite of the usefulness of deformation-mechanism maps, especially those in which experimental variables are normalized with respect to fundamental material constants such as melting-point, mechanical moduli and Burgers vectors, their use seems to have fallen off somewhat since the original enthusiasm. They are also scattered over a wide range of individual papers. It is hoped that the compendium which makes up the remainder of this work will serve as a useful reference work and also encourage the more extensive use of deformation-mechanism maps.

Aluminium-

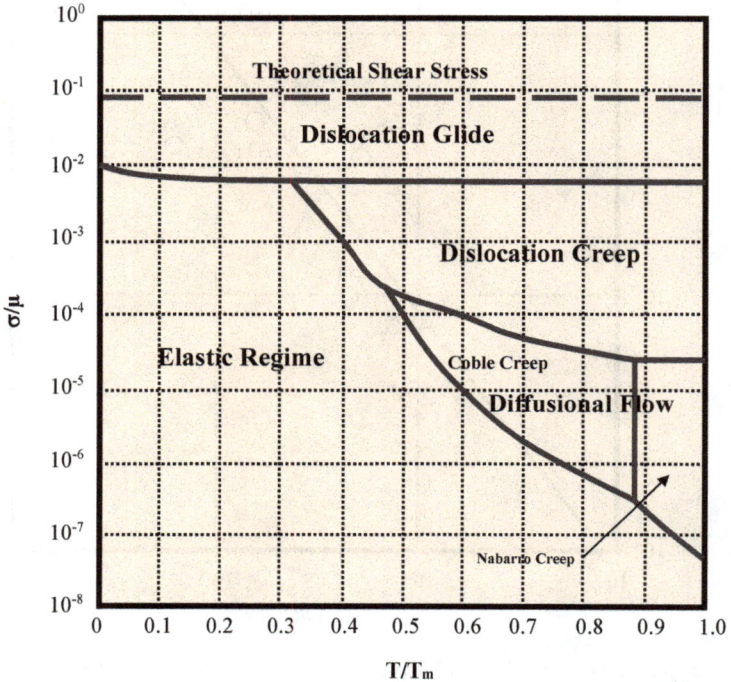

*Figure 2. Deformation mechanism map for aluminium,
with a grain size of 32 μm, at a critical strain rate of 10^{-8}/s*

An early map for aluminium (figure 2) pointed out[6] the Nabarro and Coble creep fields and the particularly large dislocation-creep field. In calculating the boundaries, the atomic volume of aluminium was taken to be 1.66×10^{-23} cm^3, the Burgers vector was 2.86×10^{-8} cm, the room-temperature shear modulus was 2.54×10^{11} dyne/cm^2 and the temperature-dependence of the shear modulus was 5.4×10^{-4}/K. In addition, the bulk and grain-boundary diffusivities were given by

$$D_l(cm^2/s) = 0.035\exp[-28.8(kcal/mol)/RT]$$

$$D_{gb}(cm^2/s) = 0.1\exp[-14.4(kcal/mol)/RT]$$

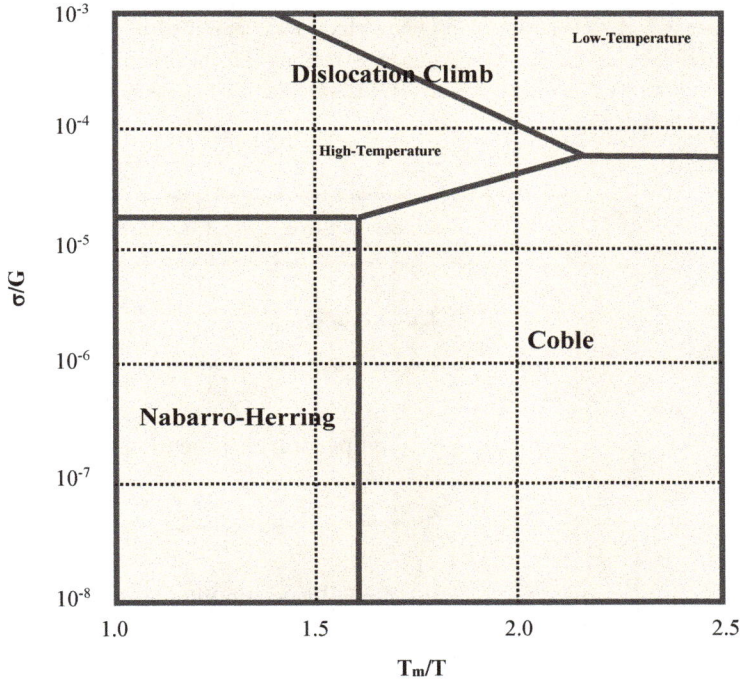

Figure 3. Deformation-mechanism map for aluminium with a grain size of 100μm

A later map[7] (figure 3) demonstrates the beneficial effect of using the reciprocal homologous temperature; the field boundaries are now completely straight.

Later work[8] has considered the effect of stress relaxation in thin films rather than in bulk material. The relaxation mechanisms of internal stress in thin films were analyzed by using deformation-mechanism maps (figures 4 to 6) and incorporated features which were associated with structure, phase instability and size effects. In particular, the dependence of stress upon temperature during thermal cycling *in vacuo* was addressed. Defectless flow was here taken to refer to flow which did not depend upon the structural defect density in the specimen.

At homologous temperatures greater than 0.2, and under a sufficiently high tensile stress, stress relaxation in the films occurred mainly via dislocation climb and the strain-rates were greater than 10^{-8}/s. Dislocation climb which was controlled by pipe self-diffusion led to stress relaxation in grain-boundary regions of the film. High-temperature dislocation climb was due to lattice self-diffusion controlled stress relaxation within the grains of the film.

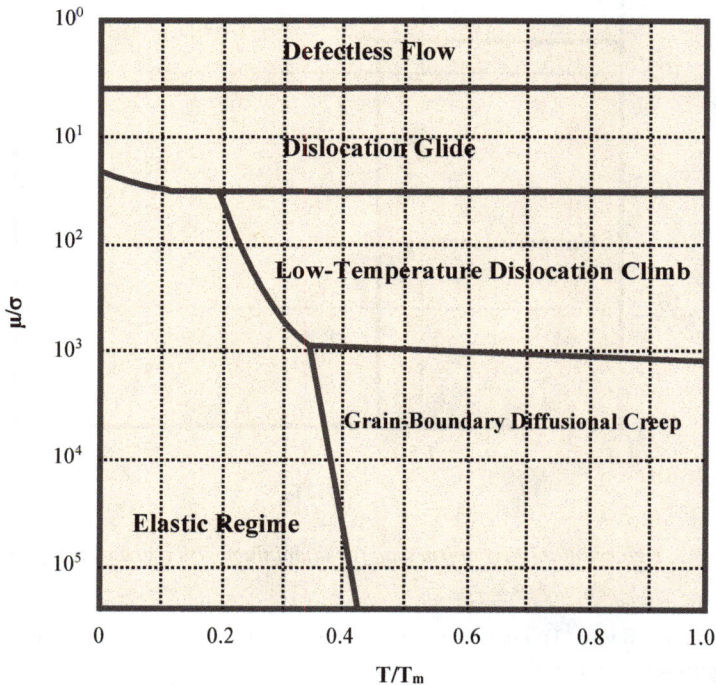

Figure 4. Deformation mechanism map for thin (0.2μm) aluminium films with grain-boundary strengthening

Two stages of stress relaxation in the films could be distinguished. In the first stage, relaxation proceeded via the most rapid mechanism. When the stress became sufficiently small, a second relaxation stage began which was controlled by the slowest plastic deformation mechanism. Situations were possible in which the second stage never occurred; for example, if the most rapid deformation mechanism was associated with stress relaxation within grains.

The effect of dislocations upon stress relaxation permitted the hysteresis in stress-versus-temperature curves to be explained in terms of the Bauschinger effect. Reference to deformation-mechanism maps gave views of the low-temperature and high-temperature yield limits which corresponded to stress relaxation in the grain-boundary regions and within the film grains, respectively.

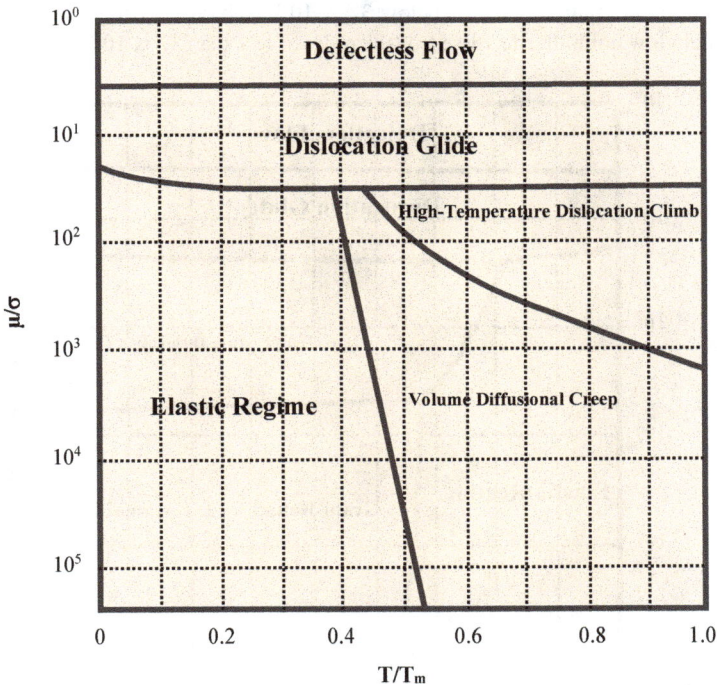

Figure 5. Deformation mechanism map for thin (0.2μm) aluminium films with essentially zero grain-boundary strengthening

The influence of grain boundary sliding upon the nature of deformation maps was considered[9]. Phenomenological equations for grain boundary sliding, which could lead to fine-structure superplastic flow, were used to show how they modify the Ashby and Langdon types of deformation-mechanism map. The former type of map was reconsidered for the modulus-normalized stress versus homologous temperature plot at a given grain size. The Langdon-type map was reconsidered for normalized grain-size versus reciprocal homologous temperature at a high constant normalized stress. Aluminium was chosen because of the earlier work, but the general conclusions were deemed to be valid for any polycrystalline solid. Grain-boundary sliding had been shown to be an important factor in the construction of deformation-mechanism maps, with large fractions of the maps commonly being dominated by grain-boundary sliding. Coble creep existed only at normalized stresses below 3.5×10^{-4}. Nabarro-Herring creep was limited to extremely low normalized stresses; of the order of less than 3.7×10^{-6}.

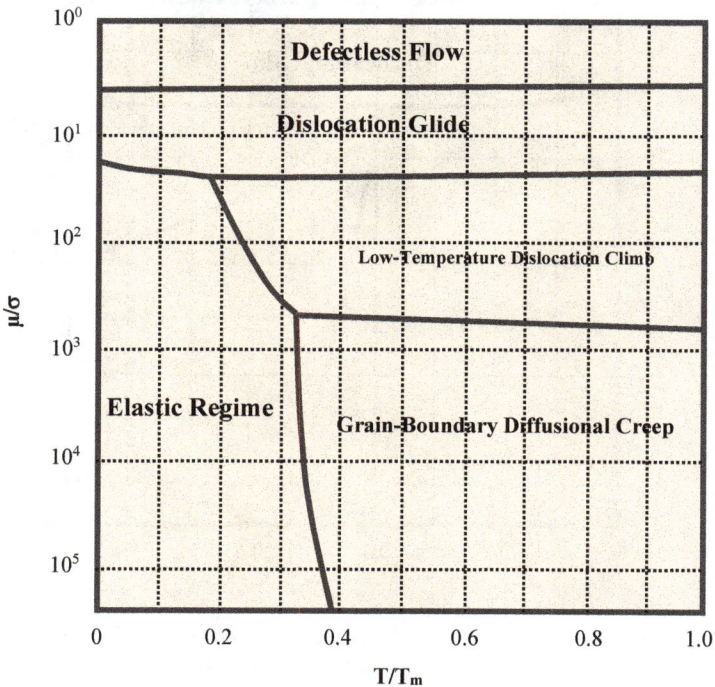

Figure 6. Deformation mechanism map for thin (0.1 μm) aluminium films with grain-boundary strengthening

It was noted that, when constructing maps with d/b plotted against the inverse homologous temperature (figures 7 and 8), the choice of the normalized stress was very important. At a normalized stress of about 5×10^{-4}, and with a grain size as large as 2μm, grain-boundary sliding would dominate deformation at all temperatures and diffusional creep would not contribute to creep.

Creep was investigated[10] in typical face-centered cubic metals at homologous temperatures below 0.3. In the case of 5N-purity aluminium, transmission electron micrographs were taken following creep testing in order to identify the intragranular deformation mechanism. All of the samples exhibited a marked creep behavior at the low temperatures, with an apparent activation energy of 15 to 30kJ/mol and a stress exponent of 2 to 5. The generation of dislocations within a cell appeared to be an enabling intragranular mechanism.

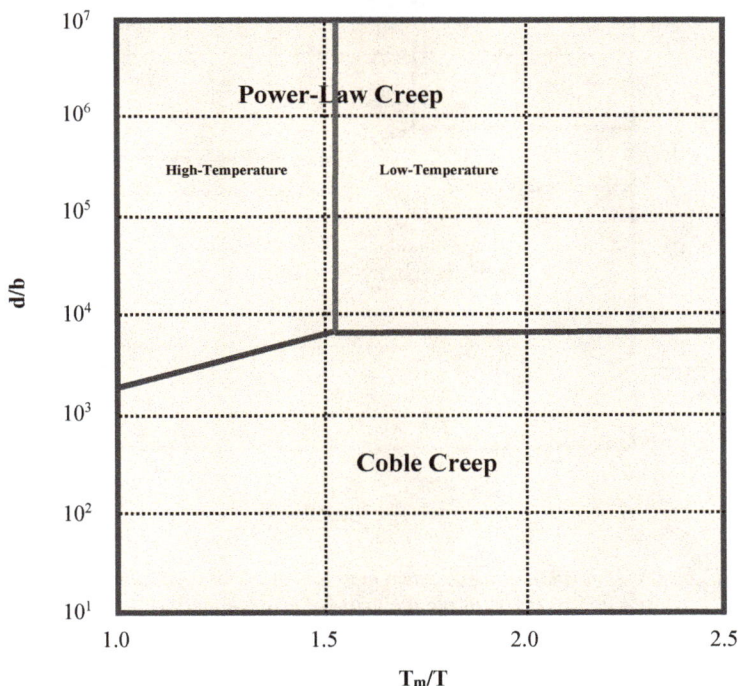

Figure 7. Deformation mechanism map for aluminium at a normalized stress of 5×10^{-4}, as originally conceived

The results revealed a new creep region that had not appeared in other deformation-mechanism maps for pure face-centered cubic metals. The original Langdon-type deformation-mechanism map was re-drawn in order to incorporate the new insight (figures 9 and 10). In the original map, the low-temperature dislocation-creep region was widely spread at low temperatures. The re-drawn map included the new creep region at low temperatures.

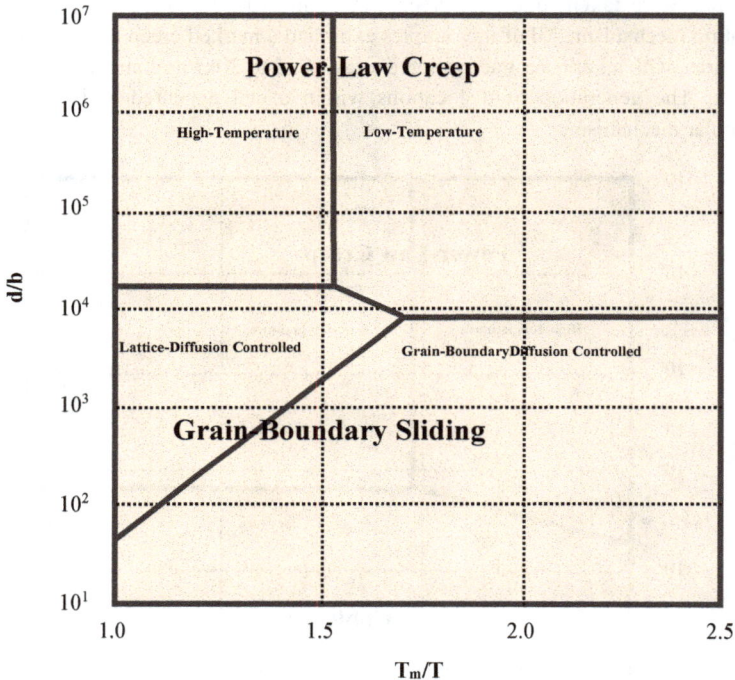

Figure 8. Deformation mechanism map for aluminium, at a normalized stress of 5 x 10⁻⁴, with the inclusion of grain-boundary sliding

It was noted[11] that deformation-mechanism maps, which had originally been developed for describing the high-temperature creep of materials having coarse grains (figure 11), could be adapted to the description of nanostructured materials. A representative deformation-mechanism map was prepared for ultrafine-grained alloy which had been subjected to equal-channel angular pressing or high-pressure torsion. Processing using equal-channel angular pressing and high-pressure torsion reduce the grains of bulk solids to sizes which are typically in the sub-micron range: 0.1 to 1.0μm.

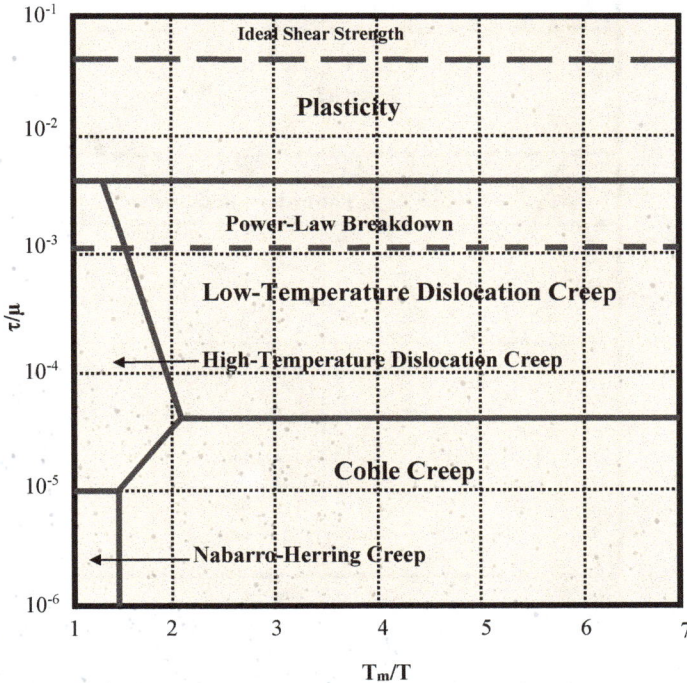

*Figure 9. Original deformation mechanism map for
5N aluminium with a grain size of 140μm*

Because the processed samples are relatively large, it was easy to test them in tension at high temperatures. Observations of ultrafine-grained materials reveal the occurrence of superplastic elongations, with the ductilities being up to 1000%. It was therefore important to produce deformation-mechanism maps which are able to depict such properties. A theoretical model for superplastic flow in large-grained materials has also been found to describe superplastic flow in ultrafine-grained materials. Using conventional creep mechanisms, maps could be constructed which described materials which had been processed using equal-channel angular pressing and high-pressure torsion.

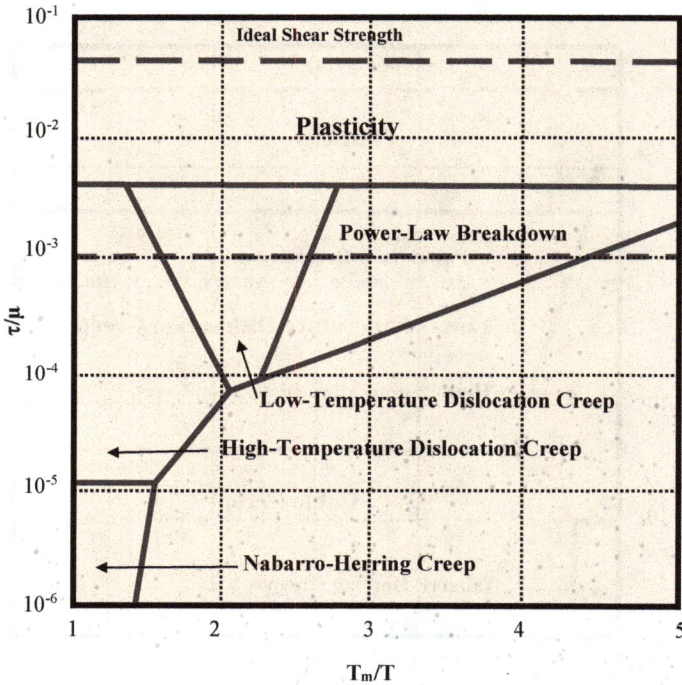

*Figure 10. New deformation mechanism map for
5N aluminium with a grain size of 140μm*

Further extension of this approach to materials having grain sizes in the nanometre range requires a more detailed understanding of the deformation mechanisms which occur at such small grain sizes. There was evidence of changes occurring in normal flow processes when the grain size was in the nanometre range. Partial dislocations might be emitted from the grain boundaries of these materials, leading to the formation of deformation twins. A critical stress may control the production of nano-twinned microstructures. The nanostructured materials may also contain non-equilibrium grain boundaries containing an excess of extrinsic dislocations. It was possible that grain-boundary sliding could become an important flow process when the grain size was greatly reduced.

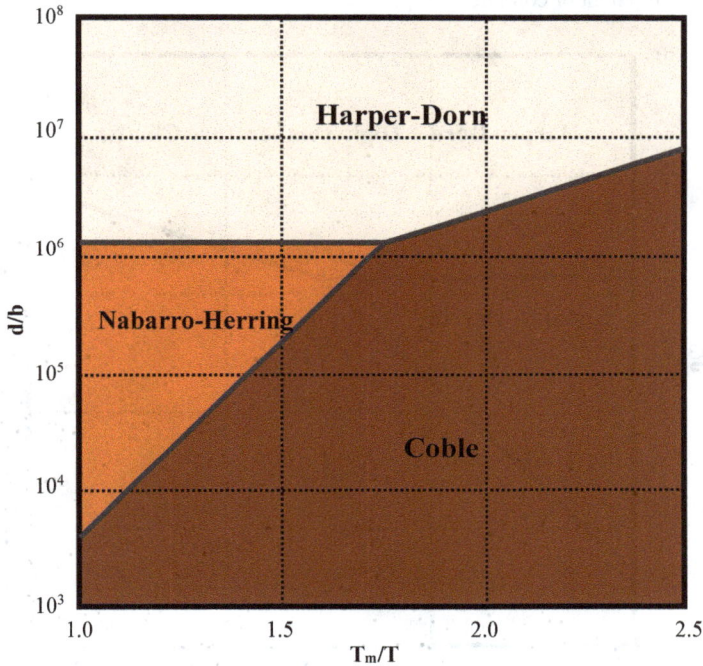

Figure 11. Deformation mechanism map for pure aluminium at $\sigma/G = 10^{-6}$

The then new type of deformation-mechanism map in which the normalized grain size was plotted against the reciprocal of the homologous temperature, was ideal for the analysis of high-temperature creep data and straightened the boundaries which separated the various fields (figures 12 and 13). It was easy to construct and offered the advantage that the basic form of the map was invariant with respect to stress for mechanisms which were governed by the same stress exponent. It was noted[12] that, under creep conditions, pure metals and solid-solution alloys usually deformed via Nabarro-Herring and Coble diffusional creep, and by dislocation-climb which was controlled by either core or lattice diffusion, at low and high temperatures, respectively. The status of Harper-Dorn creep was not fully understood at that time. There were then also few unambiguous reports of Coble creep in pure metals because the very slow creep rates required high-sensitivity measuring equipment or complicated specimens.

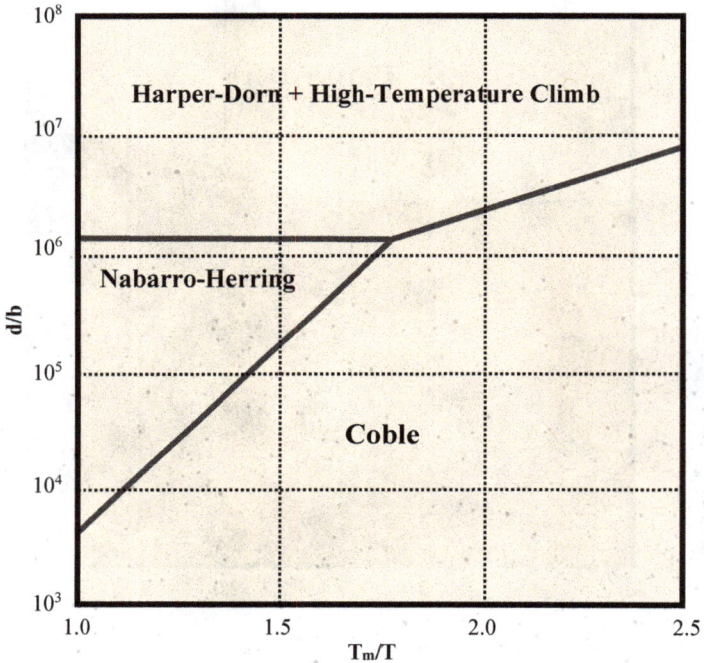

Figure 12. Deformation mechanism map for pure aluminium at $\sigma/G = 8.9 \times 10^{-6}$

Burgers vector distributions and dislocation line directions were determined[13] as a function of strain and strain-rate, showing that the Burgers vector distributions were similar at superplastic and higher strain-rates, although there was a gradual increase in the dislocation density and number of climbing dislocations as the strain-rate increased. There was no sharp change in dislocation structure between the superplastic and high strain-rate regions. Following deformation at the minimum strain-rate, the Burgers vector distribution was uneven and insufficient dislocations were available to maintain grain compatibility by slip alone. It was concluded that diffusion processes then became more important. The present map was constructed for pure aluminium with a grain size of

3μm. Specimens which were deformed at the highest strain-rate were in a dislocation creep region, while samples which were tested at lower strain-rates were in an area where both dislocation motion and grain-boundary diffusion were important. The diffusion paths for volume diffusion could be very short, and volume diffusion could therefore be important at low strain-rates, in spite of the relatively low temperature.

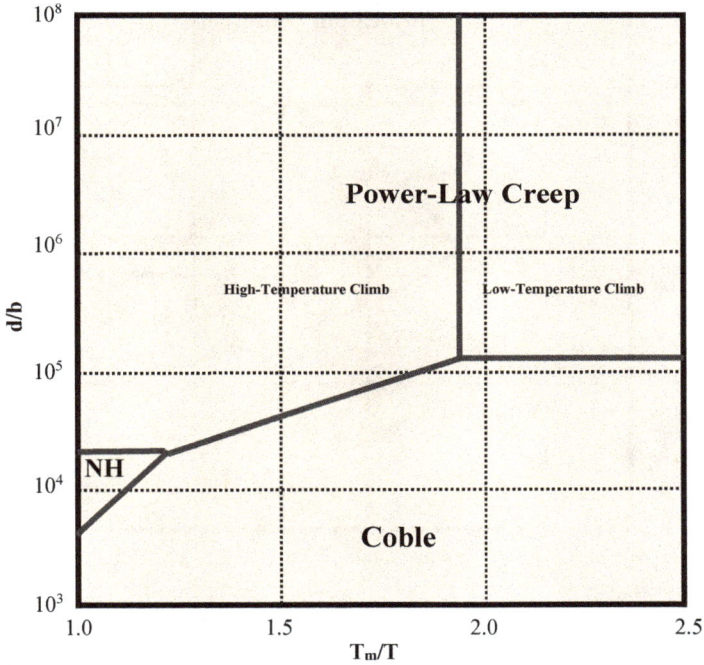

Figure 13. Deformation mechanism map for pure aluminium at $\sigma/G = 10^{-4}$

Two mechanisms, dislocation motion and diffusional flow, acted in series with grain-boundary sliding to accommodate this deformation mode (figure 14). The faster of these two mechanisms was rate-controlling. The fact that dislocations were created and mobile, and the similarity of the Burgers vector distributions except at the lowest strain-rate, implied that a dislocation process was the rate-controlling mechanism; at least near to the high strain-rate limit of the superplastic region. The observations also implied that dislocation motion acted as an independent deformation mechanism during superplastic flow. The climb of dislocations through grains or along grain boundaries, and subsequent annihilation, were thought to be rate-controlling accommodation mechanisms.

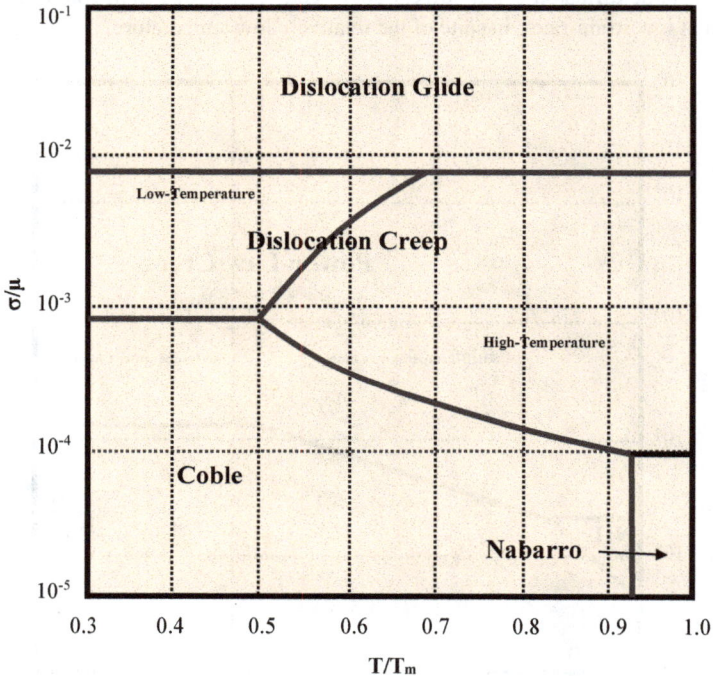

Figure 14. Deformation mechanism map for aluminium with a grain size of 3μm

Maps were developed[14] for pure aluminum at two different homogeneous temperatures, and the relative contributions of the various deformation processes were estimated as a function of stress at differing grain sizes. The maps could be easily constructed from a bare knowledge of the relevant constitutive equations for the various mechanisms. At the time, there was considerable evidence that pure aluminium deformed mainly via some form of dislocation climb process at high temperatures and relatively high stresses. Harper-Dorn creep was expected to occur at very low stresses and at temperatures near to the melting-point. The creep of aluminium having a large grain size occurred via Harper-

A Compendium of Deformation-Mechanism Maps for Metals   Materials Research Forum LLC
Materials Research Foundations **116** (2022)   https://doi.org/10.21741/9781644901694

Dorn creep. At low stress levels and small grain sizes, plastic deformation usually occurred via the diffusion of vacancies through the crystalline lattice (Nabarro-Herring) or along the grain boundaries (Coble). Noting that each of the mechanisms operated independently, and that the various individual strain-rates were additive, predicted the observed creep rate. Deformation-mechanism maps were constructed (figures 15 and 16) for homologous temperatures of 0.5 and 0.9. The d/b values ranging from 10^3 to 10^8 correspond to grain sizes of $0.3\mu m$ to 3cm and therefore cover all of the grain sizes which are likely to be met with in pure aluminium.

The processing of bulk metal using extreme plastic deformation leads to grain refinement, with the grain sizes typically being in the sub-micron or nanometre range. If the small grains are suitably stable at high temperatures, the ultrafine-grained metals exhibit marked superplastic properties in tension at high temperatures. Ultrafine-grained materials are commonly produced using equal-channel angular pressing or high-pressure torsion. This work demonstrated the feasibility of preparing deformation mechanism maps which could provide comprehensive information concerning the flow mechanisms. Considering Al-33%Cu alloy, tested at 723K[15], the results (figure 17) could be divided into 3 distinct regions of flow. Region II, at intermediate strain-rates was the true superplastic region, with elongations-to-failure of up to more than 2000%. Region III, at high strain-rates, was due to a transition to a dislocation mechanism such as dislocation climb. Region I, at low stresses, arose from the segregation of impurity atoms at grain boundaries.

This map (figure 18) was the first demonstration[16] of the possibility, mentioned in the introduction, that two processes might interact in a sequential manner. The procedure which was used in this study was of general utility validity and could be used for a wide range of situations in which two mechanisms were known to operate sequentially.

There are specific ranges of stress and grain size in the present map, at an homologous temperature of 0.9, within which each of 5 possible deformation mechanisms may be the rate-controlling process.

Climb: strain-rate $\propto (D_lGb/kT)(\sigma/G)^4$

Glide: strain-rate $\propto (D_lGb/kT)(\sigma/G)^3$

Harper-Dorn: strain-rate $\propto (D_lGb/kT)(\sigma/G)$

Nabarro-Herring: strain-rate $\propto (D_lGb/kT)(b/d)^2(\sigma/G)$

Coble: strain-rate $\propto (D_{gb}Gb/kT)(b/d)^3(\sigma/G)$

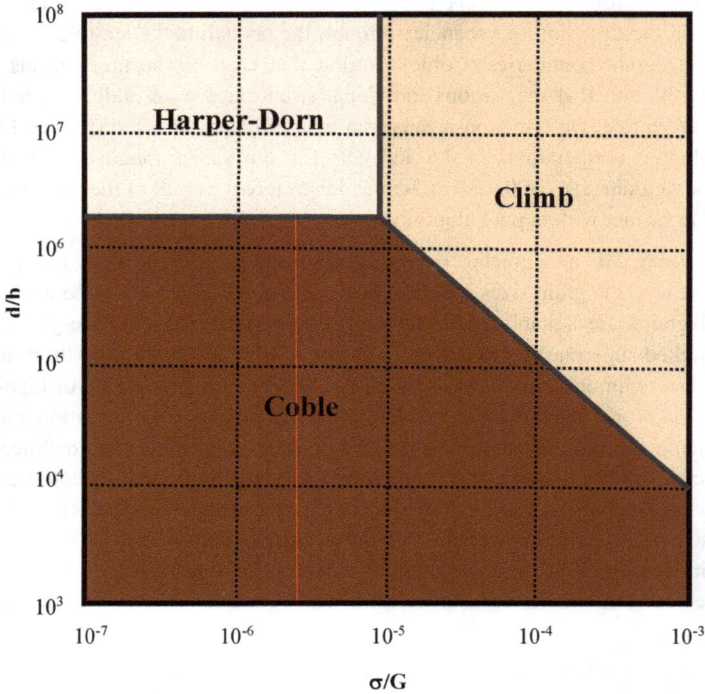

Figure 15. Deformation mechanism map for pure aluminium at an homologous temperature of 0.5

On the other hand, because too small a grain size is unstable at this temperature, the field of existence of Coble creep is not of much significance in practice. Strain arises from glide in both the glide and the climb fields, and the division between these fields depends only upon the relative magnitudes of the rates of climb and glide. The slope of the line between any 2 fields is known to be governed by the difference in stress exponent, divided by the difference in inverse grain-size exponent for the mechanisms on either side of that line. There is therefore a slight change in slope in going from the boundary between climb and Nabarro-Herring behavior to the boundary between glide and Nabarro-Herring behavior.

In general, the value of d/b which separates Nabarro-Herring and Coble fields decreases, with increasing temperature, at a rate which is three times more rapid than the transition value for Harper-Dorn and Coble creep. The boundary between Harper-Dorn and Coble may be eliminated if it falls within the field of Nabarro-Herring creep. For a given homologous temperature, the rate-controlling mechanisms acting under any conditions of stress and grain-size may be known but do not reveal the relative contributions made by the various mechanisms at a given point.

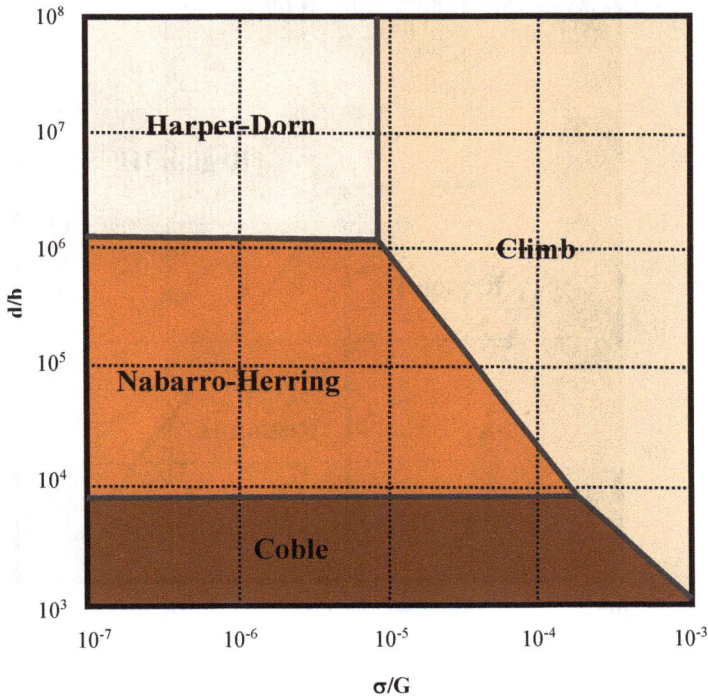

Figure 16. Deformation mechanism map for pure aluminium at an homologous temperature of 0.9

Thermal cycling of a sputter-deposited thin film of Al-1mol%Si alloy on a silicon substrate was carried out[17] between room temperature and 723K. The residual stress increased with increasing cycles, up to the fourth. There was then a continuous decrease with further cycling. The initial increase in residual stress was attributed to an increase in lattice dislocations and to tangling. The subsequent decrease in residual stress was due to crack formation and delamination. Stress relaxation was monitored during isothermal annealing at various temperatures, and analysis of the relaxation curves revealed three temperature regions which reflected various deformation mechanisms. The boundaries between the neighboring regions corresponded to the boundaries in a deformation-mechanism map for a thin film of aluminium (figure 19).

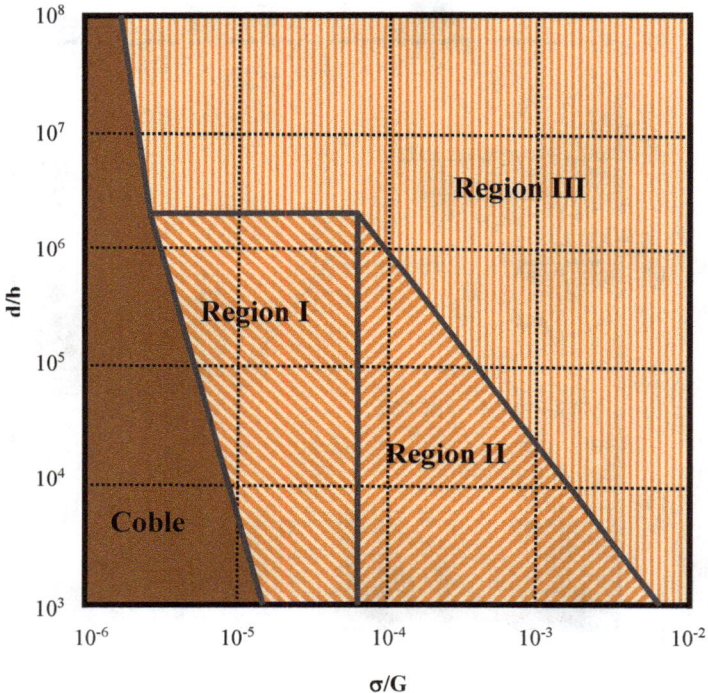

Figure 17. Deformation mechanism map for Al-33%Cu at 723K. Regions I, II and III denote the plastic flow associated with superplastic metals in sigmoidal, Nabarro-Herring and Coble creep

The superplastic behavior of Al-12.7Si-0.7wt%Mg was investigated[18], and reasonable elongations were found, for samples having a 9.1μm grain size, at 733 to 793K, under strain-rates of 1.67×10^{-4} to 1.67×10^{-3}/s. A maximum elongation-to-fracture of 379%

was found, together with a strain-rate sensitivity of 0.52 and an activation energy for flow of 156.7kJ/mol at 793K and an initial strain-rate of 1.67 x 10^{-4}/s. This value was close to the activation energy for lattice diffusion in aluminium. Dislocation observations made in aluminium grains showed that intragranular slip was the accommodation mechanism involved in grain-boundary sliding. Most of the grain boundaries were high-angle, indicating that boundary sliding and grain rotation occurred during deformation.

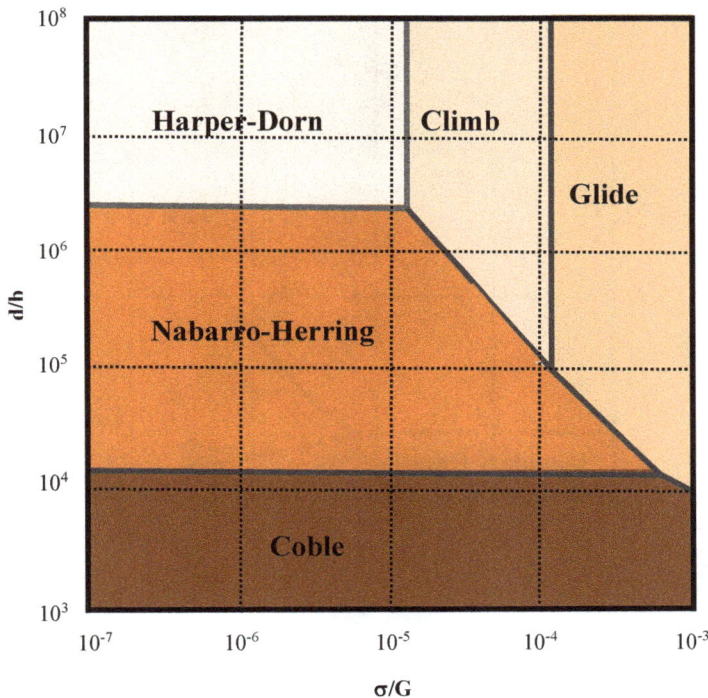

*Figure 18. Deformation mechanism map for Al-3%Mg
at a homologous temperature of 0.9*

The deformation mechanism map (figure 20) showed that the experimental data were in very close agreement with its predictions. Most cavities formed around silicon particles, suggesting the operation of a cavity-formation mechanism. Filaments were observed on fracture surfaces, and their quantity or density increased with increasing test temperature.

This was attributed to a transition from dislocation viscous glide creep to grain-boundary sliding at high temperatures. The formation of filaments was related to the deformation mechanisms and to the lattice diffusion at high temperatures.

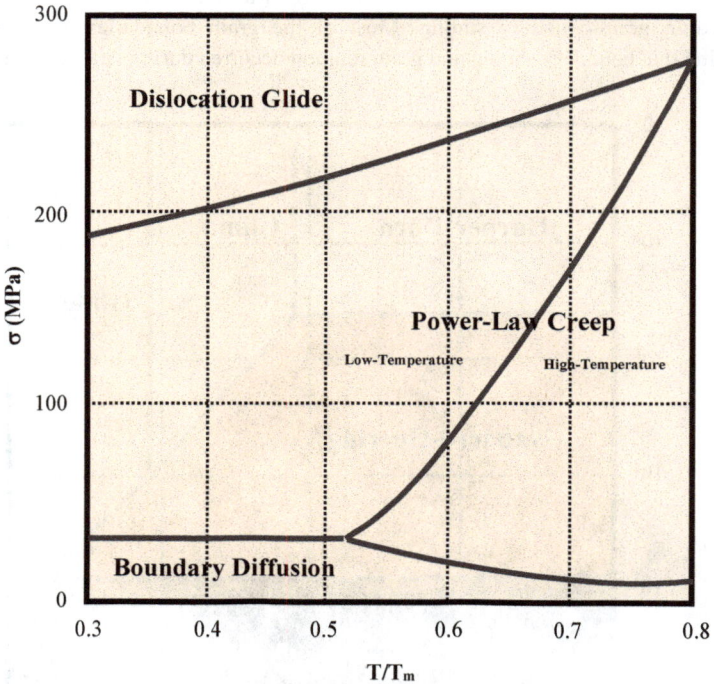

Figure 19. Calculated deformation mechanism map for aluminium thin film on silicon

Plastic deformation of sub-µm sized metals at different temperatures is influenced by factors which are absent from their bulk counterparts, including surface diffusion assisted softening and mechanical/thermal annealing-induced hardening. The test temperature and sample size therefore strongly affect the mechanical behavior, necessitating the construction of new deformation-mechanism maps. Here, based upon results from *in situ* quantitative compression tests on micro-pillars at various sizes and temperatures ranging up to 400C, maps were constructed[19] for monocrystalline sub-micron scale aluminium, consisting of elasticity, diffusive plasticity and displacive plasticity regimes. In the sample-size versus stress map (for a fixed temperature), a so-called strongest size was found at the triple junction of three regimes; above which smaller was stronger, and below which smaller was weaker. In the diffusive plasticity regime, deformation is localized within the top pillar volume and demarcated by a moving-front interface, which is likely to be a newly formed grain boundary that is impenetrable to impinging dislocations below a critical stress of $\sim 1\,\mathrm{GPa}$.

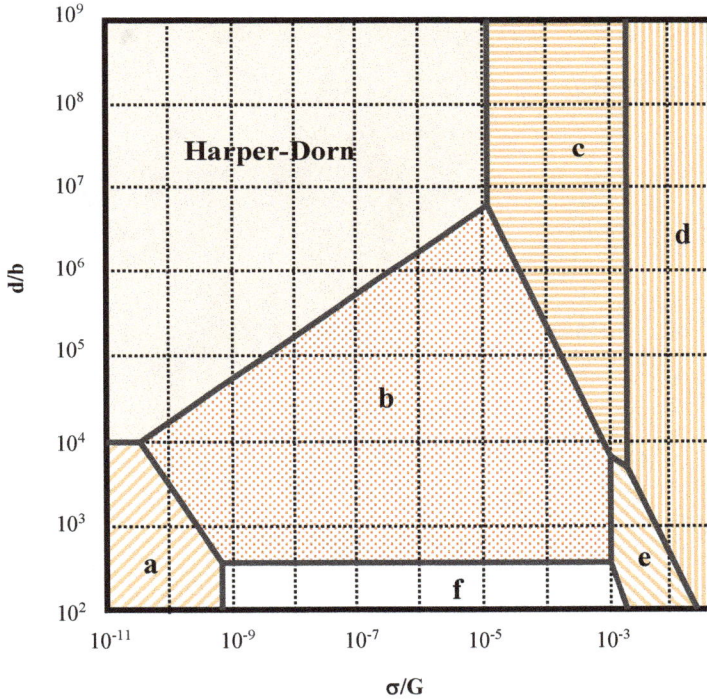

Figure 20. Deformation mechanism map for Al–Si alloy at 793K; a – diffusional flow controlled by grain-boundary diffusion and a stress exponent of 1, b – superplastic grain-boundary sliding controlled by lattice diffusion and a stress exponent of 2, c – dislocation slip controlled by lattice-diffusion and a stress exponent of 5, d – dislocation slip controlled by lattice-diffusion and a stress exponent of 7, e – dislocation pipe grain-boundary diffusion and a stress exponent of 4, superplastic grain-boundary sliding controlled by grain-boundary sliding and a stress exponent of 2

The hot-deformation behavior of Al–0.2Sc–0.04wt%Zr was investigated[20] by conducting hot-compression tests at temperatures of 440 to 600C and at strain-rates ranging from 0.001 to 5/s. The characteristics of the true stresses and strains which were acquired in the hot-compression tests were investigated, and the influence of processing parameters upon the microstructure of the deformed samples was observed by using optical microscopy and electron back-scattered diffraction.

Cadmium

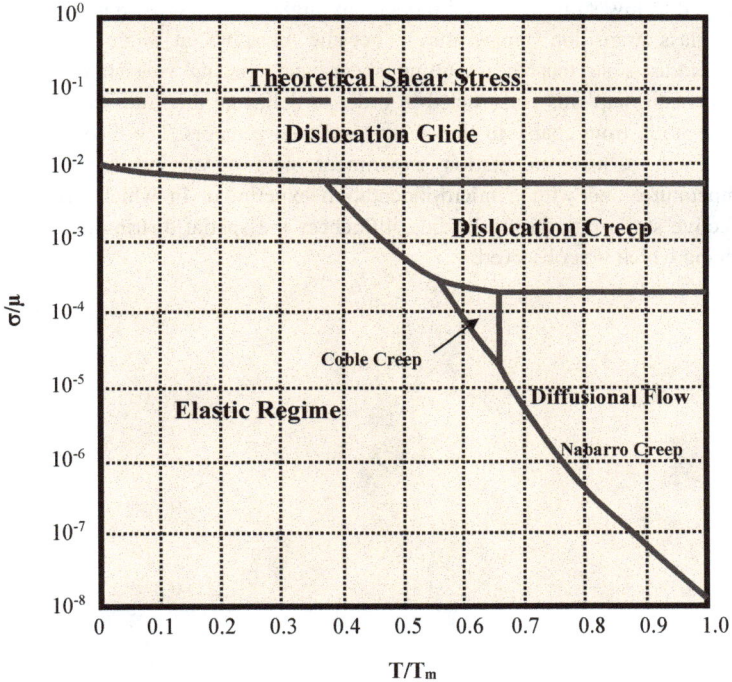

Figure 21. Deformation mechanism map for cadmium,
with a grain size of 32μm, at a critical strain rate of 10^{-8}/s

Recent experiments have shown that inhomogeneous deformation in amorphous alloys depends critically upon the environmental temperature and the applied strain rate, and that the temperature field inside the shear-band can attain the glass transition temperature[21]. A free-volume based, thermo-viscoplastic constitutive law was developed[22] in which the thermal transport equation included contributions arising from heating due to plastic work and from heat conduction. For homogeneous deformation, the instantaneous temperature rise during the strain softening stage could lead to thermal softening and promote the initiation of shear bands. A linear stability analysis was carried out in order to examine the conditions for the unstable growth of temperature fluctuations. It was found that short-wavelength fluctuations, the amplitudes of which decayed at low strain-rates and moderately high temperatures (but still much lower than the glass transition temperature), became unstable at high strain-rates and low temperatures, so that the resulting shear-band spacing was shorter. A deformation-mechanism map was constructed in order to delineate this transition in inhomogeneous deformation from coarse to fine shear-band arrangements. The theoretical results agreed well with a nano-indentation experiment with a varying applied strain-rate and temperature, and with a micro-indentation experiment in which the evolution of the effective strain-rate during loading influences the spatial distribution of the shear-band spacing which was observed.

Cobalt-

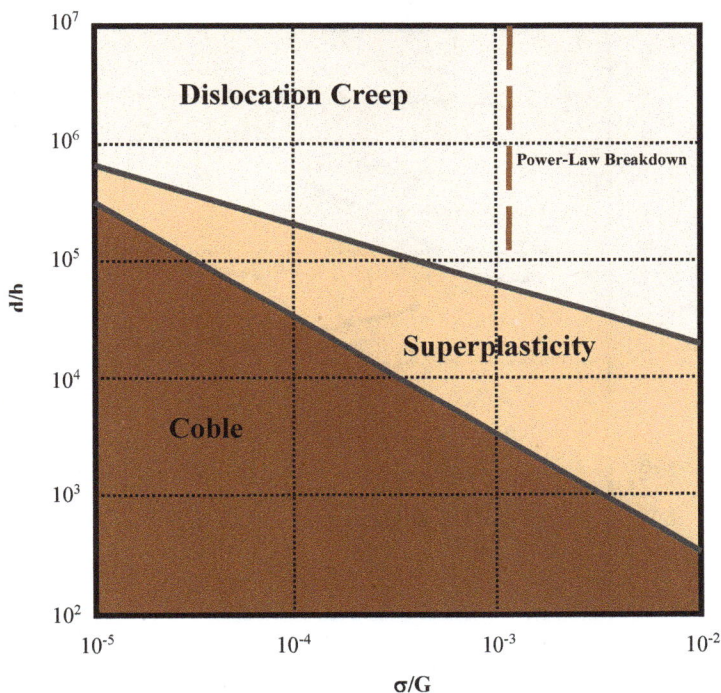

Figure 22. Deformation mechanism map for CoCrFeMnNi alloy at 1023K

This was an initial attempt[23] to develop a deformation-mechanism map for a member of the new class of high-concentration solid solutions known as high-entropy alloys. It was found that diffusion was not appreciably slower in such alloys, and that superplastic deformation adhered to the standard model deduced for conventional polycrystals. The data suggested that deformation at higher stresses occurred via dislocation glide creep, followed by a power-law breakdown regime. It was necessary to be cautious when interpreting spherical nano-indentation creep data, but some results on nanocrystalline samples suggested that room-temperature deformation occurred via Coble diffusion creep.

Copper-

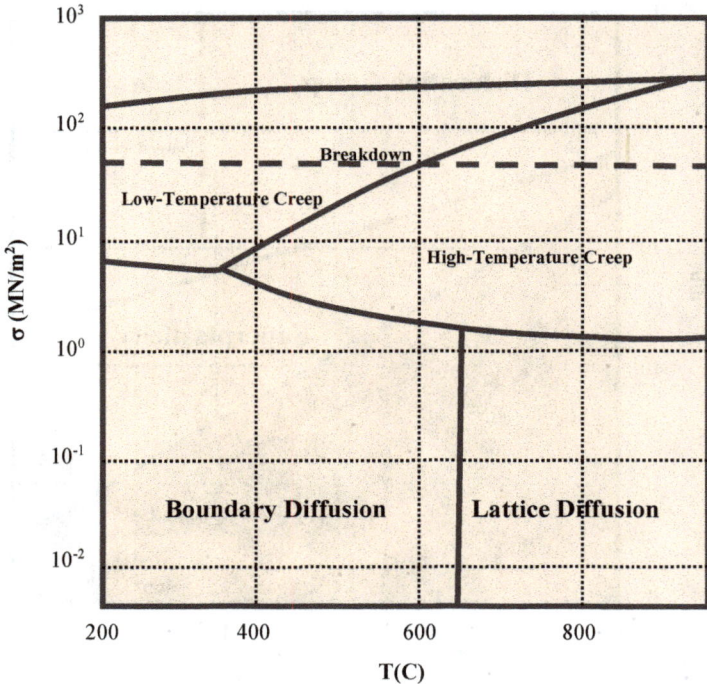

Figure 23. Deformation mechanism map for polycrystalline copper

A new approach to creep studies was illustrated[24] with regard to polycrystalline copper. The complex stress- and temperature-dependences of the minimum creep-rate and the rupture-life, together with the exact form of the deformation-mechanism map could be predicted directly.

Cold-spray processing is a solid-state coating technique and an emerging method for additive manufacturing, in which metal powder particles are bonded through high-velocity impact-induced deformation. However, the severe plastic deformation of powder particles at extremely high strain rates, high strain gradients, and localized elevated temperatures yields rather complex and heterogeneous microstructures in the materials produced in cold sprayed coating or bulk forms[25]. A good understanding, and even prediction, of such heterogeneous microstructures is essential for determining suitable post-processing conditions as well as the final properties of cold sprayed products. A cold-spray system was used to deposit copper coatings over a temperature range of 373 to 873K, and a systematic investigation was made of the microstructural evolution of the coatings using electron back-scatter diffraction and transmission electron microscopy.

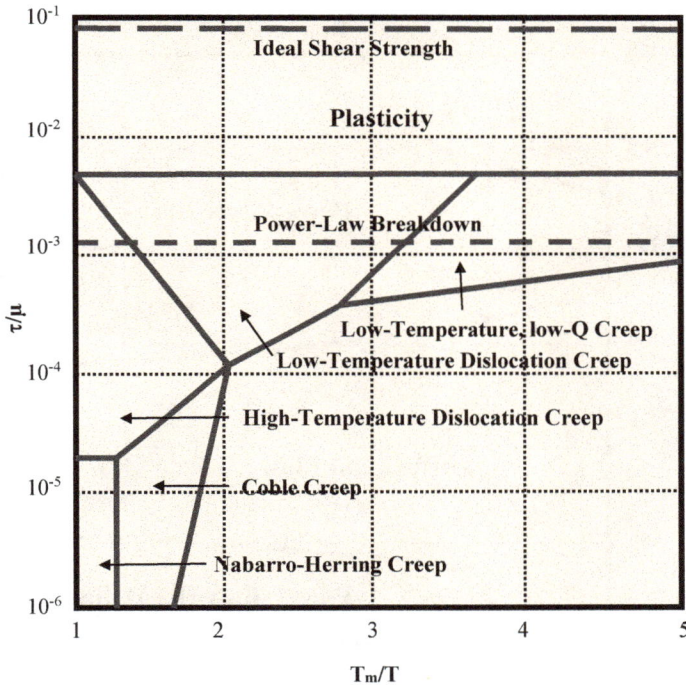

Figure 24. New deformation mechanism map for 4N copper with a grain size of 44μm

Diverse microstructures were observed[26], including recrystallized grains, annealing twins, shear bands, sub-micron grains, deformation twins and nanometre-sized grains. In order to understand the formation of such complex microstructures, the local strains, strain-rates and temperatures of the cold-sprayed powder particles were deduced by using finite-element methods. Based upon experimental and simulation results, a first deformation mechanism map was created for cold-sprayed coatings so as to interpret and predict heterogeneous microstructural evolution in copper by using the local Zener-Hollomon parameter and plastic strain. Such a map could be used to predict and design the microstructures of cold-sprayed copper samples on the basis of processing parameters and could also be extended to other severe plastic deformation processes such as cutting, extrusion, solid-phase welding and divers solid-state additive manufacturing processes.

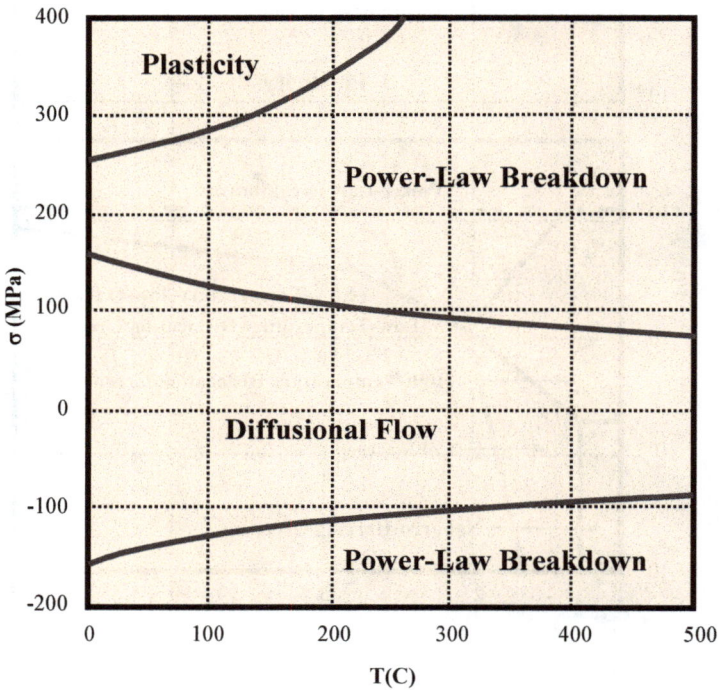

Figure 25. Deformation mechanism map for 1μm copper film with a grain size of 0.1μm on silicon

The mechanical response of *in situ* copper-chromium composite was modeled[27] by using the deformation-mechanism map approach. The stresses in each phase were predicted as a function of temperature and strain rate. From this the ratio of phase stresses, and hence the degree of load transfer, was obtained. The extent of load transfer and the predicted deformation modes of the two phases were then related to the expected failure mechanisms of the composite. Three modes of composite failure were predicted. Copper-chromium *in situ* composites, and pure copper, were produced via a casting and swaging route. The mechanical properties were then characterized experimentally by tensile testing. Mechanical tests confirmed that the composite exhibited a significantly higher strength than that of the unreinforced material. The observed failure modes of the composite were compared with the predictions.

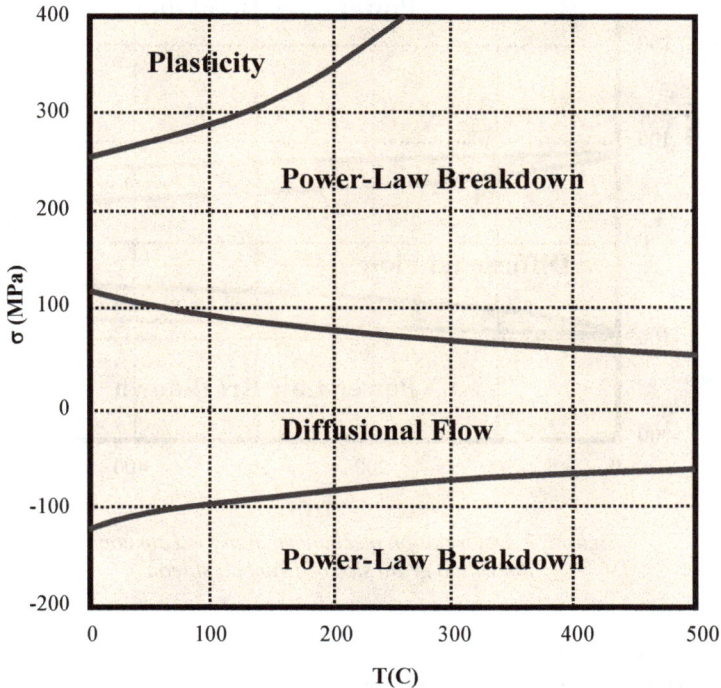

Figure 26. Deformation mechanism map for 1µm copper film with a grain size of 1µm on silicon

On the basis of existing creep models, a prediction was made[28] of the stresses developed in thin films of copper on silicon wafers that were subjected to thermal cycling. The results were plotted on deformation-mechanism maps which identified the predominant mechanism which operated during thermal cycling.

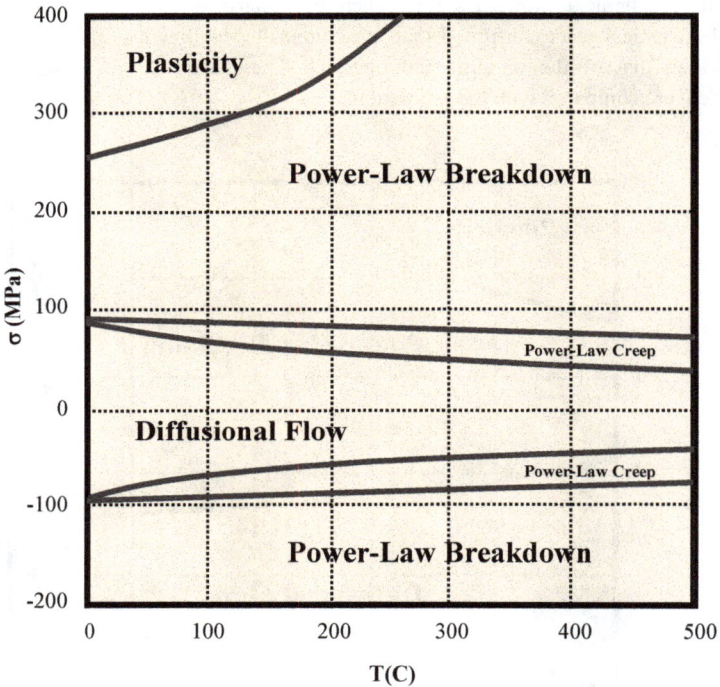

Figure 27. Deformation mechanism map for 1μm copper film with a grain size of 10μm on silicon

The predictions were compared with temperature-ramped and isothermal stress measurements of a 1μm-thick sputtered copper film at 25 to 450C. The models predicted the rate of stress relaxation when the film was held at a constant temperature, and the stress-temperature hysteresis which occurred during thermal cycling. In the case of 1μm-thick copper films which were cycled at 25 to 450C, the deformation maps indicated that grain-boundary diffusion controlled stress relief at temperatures above 300C, where only a low stress could be maintained in the films. Power-law creep was important at intermediate temperatures and determined the maximum compressive stress.

Deformation-mechanism maps were constructed[29] for CuNi thin films. In determining the plastic deformation mechanisms, use was made of the strain-rate equations for bulk material. Modifications were made for thin films. The results showed that the plastic deformation which was associated with dislocation motion was over-estimated, and that diffusive effects strongly affected the form of the map, the predominant deformation mechanism and the strain-rate. It was concluded that such maps could offer only an approximate guide to the operating plastic deformation mechanisms.

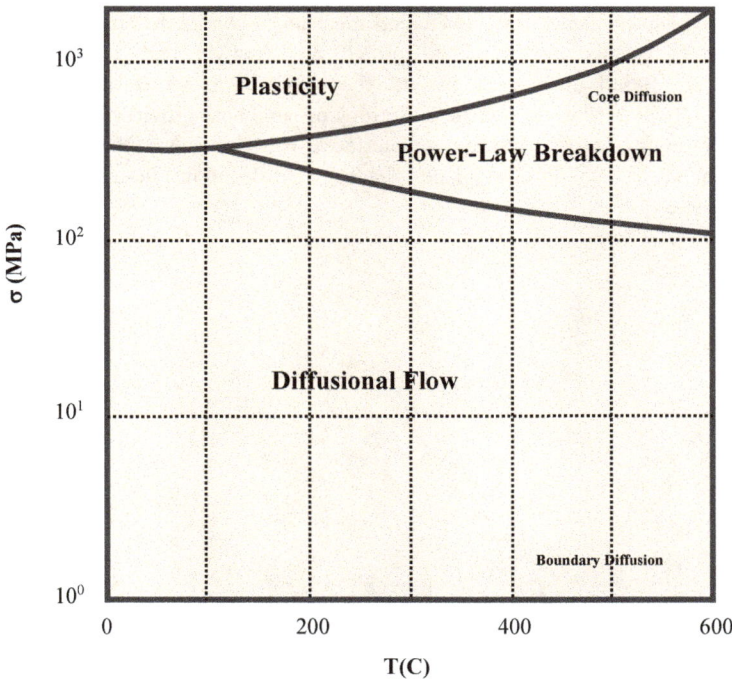

Figure 28. Deformation mechanism map, biaxial stress versus temperature, for $Cu_{0.6}Ni_{0.4}$ thin films with a thickness and grain size of 400nm

Although cavity formation had been reported in copper thin films, its mechanism had not been understood. The present work was aimed[30] at understanding the cavitation mechanism in relation to deformation and stress-concentration mechanisms. The samples consisted of layers of Cu/Ta/SiO$_2$/Si. The tantalum layer was deposited as a diffusion-barrier. Stress changes were measured in copper thin films during thermal cycling at heating and cooling rates of 0.056K/s at temperatures ranging from ambient to 723K. Morphological changes were monitored by using scanning and transmission electron microscopy. Analysis of stress-temperature curves indicated that the stress state was tensile during cooling and compressive during heating. The resultant stress-temperature curve was compared with a calculated deformation-mechanism map. It was found that deformation occurred mainly via a dislocation-glide creep mechanism during cooling, and via a grain-boundary diffusion creep mechanism during heating. Microstructural observations revealed that cavities formed at twin|twin and twin|grain-boundary intersections. Based upon these results, the cavity formation mechanism could be understood by proposing that elastic anisotropy of neighboring twin variants gave rise to the concentration of shear stress at twin interfaces. This then caused the twin interfaces to be preferential dislocation glide planes, leading to dislocation pile-up and cavitation at the intersection.

Iron-

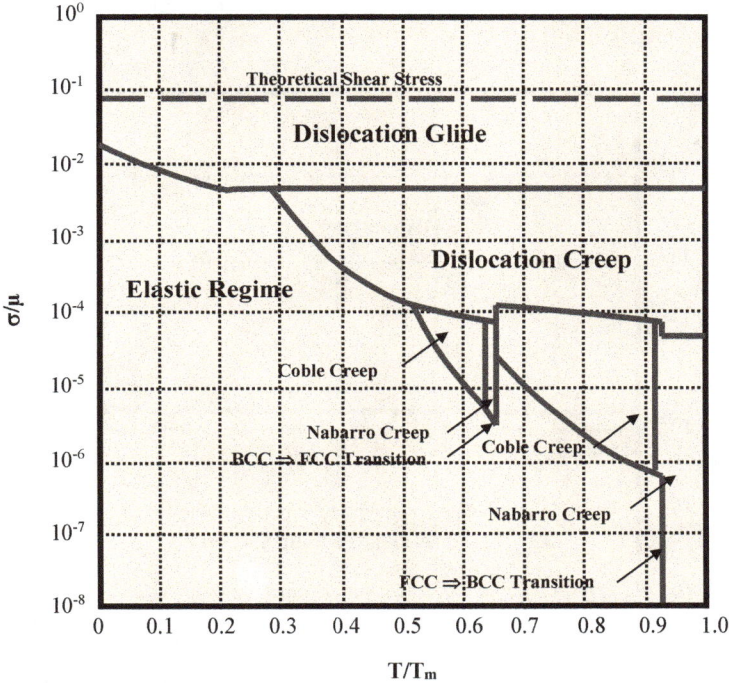

Figure 29. Deformation mechanism map for iron,
with a grain size of 32 μm, at a critical strain rate of $10^{-8}/s$

Stress-strain curves, the relationship between differential stress and macroscopic sample strain, of polycrystalline ε-iron were obtained[31] at pressures of 17GPa, three different temperatures (600, 400, 300K) and various strain rates between 3.8 x 10^{-6} and 2.3 x 10^{-5}/s. Five independent stress-strain curves were obtained[32] upon axial shortening, and the sample exhibited an overall ductile behavior. At above 4% axial strain, the sample stresses reached saturation and the sample exhibited steady-state deformation. The stress exponents at temperatures of 400 and 600K were determined to be ~31 and ~7, respectively. These results indicated that ε-iron deforms in the plastic regime when below

400 K and that the dominant deformation mechanism at 600K may be low-temperature power-law creep. The overall deformation behavior for ε-iron was consistent with that of zinc; suggesting that the deformation-mechanism map of ε-iron resembled those of other hexagonal metals.

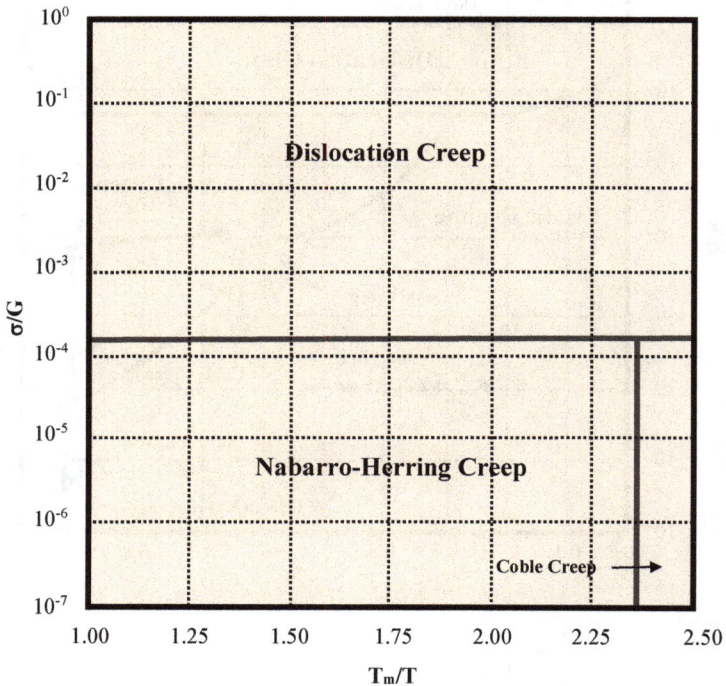

Figure 30. Deformation mechanism map for FeAl

The creep behavior of Fe-Ni-Al alloys having the intermetallic (Ni,Fe)Al phase (B2 structure) as matrix and α-Fe precipitate particles (disordered body-centered cubic structure) was studied[33] at 900 and 1000C and compared with that of single-phase B2-(Ni,Fe)Al and other two-phase Fe-Ni-Al alloys containing the (Ni,Fe)Al phase. The observed creep behavior was controlled mainly by a dislocation creep which was described by conventional power-law (Dorn equation) with a threshold stress for the strengthening precipitates. Deformation-mechanism maps were calculated. The creep resistance of the alloys was of the order of that of the nickel-based superalloy, Inconel 617.

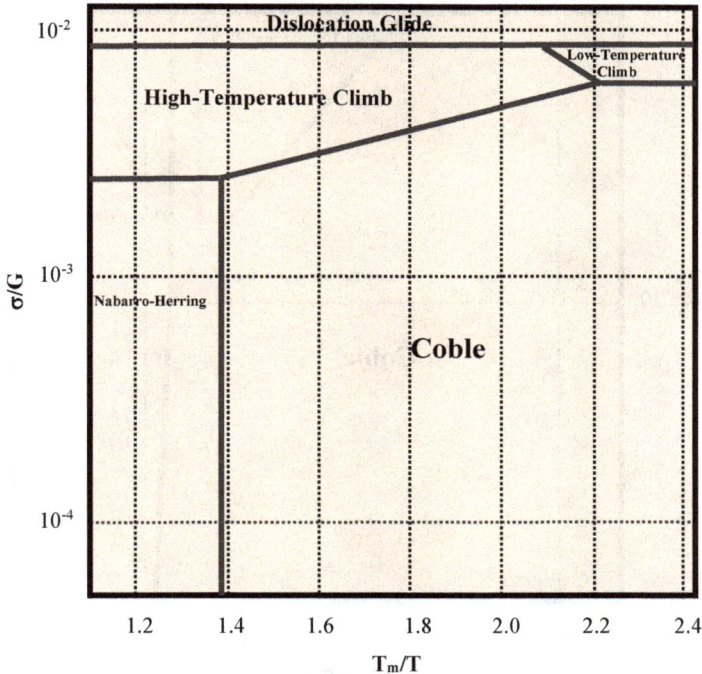

Figure 31. Deformation mechanism map for 1Cr-Mo-V steel with a grain size of 23μm

Short-term creep tests were performed[34] on a 9%Cr (P-91 type) steel at 873 to 923K, and at stresses below 100MPa, by using the helicoid-spring specimen technique. The steady-state creep rates corresponded to viscous behavior under the above conditions, and were characterized by an apparent stress exponent which was close to unity. Because the stress-exponent at higher stresses was about 10, the change in the deformation mechanism at lower stresses was evident. The deformation-mechanism map which resulted showed that the loading conditions responded to the viscous creep. Extrapolation from the power-law creep regime to low stresses could lead to a serious under-estimation of the predicted deformation rates. The primary stage could be described by Li's equation. Preliminary annealing at 873 K for 1.8 x 10^7s reduced the primary strain, but it had no effect upon the steady-state creep rate.

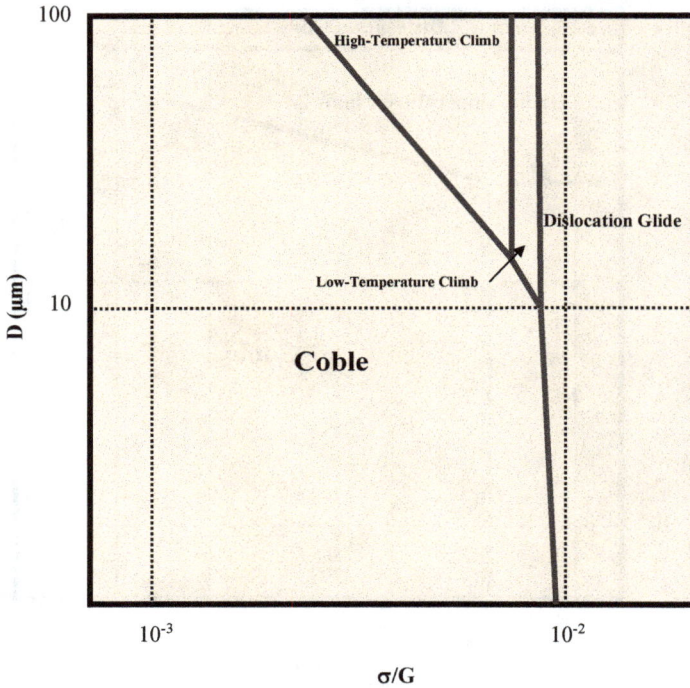

Figure 32. Deformation mechanism map for 1Cr-Mo-V steel at 565C

Various studies by different investigators had produced[35] high strain-rate superplasticity in numerous materials. These materials had, in common, ultra-fine grain sizes (1 μm and less) and maximum ductilities at high strain-rates (0.1/s and above) when deformed at elevated temperatures. A model which had been proposed for high strain-rate superplasticity when incipient melting was not present, used grain boundary sliding - accommodated by dislocation glide across sub-grains - with a rate-controlling contribution from dislocation-pipe diffusion to explain the observed behavior. Predictive equations for this and other mechanisms were used to construct deformation-mechanism maps. Data on an ultra high-carbon steel were compared with the predictions. A numerical model was used to predict the tensile ductility as a function of the stress exponent; which was controlled by the dominant deformation mechanism. At temperatures where incipient melting was not expected, these predictive relations explained the observed high strain-rate superplastic behavior.

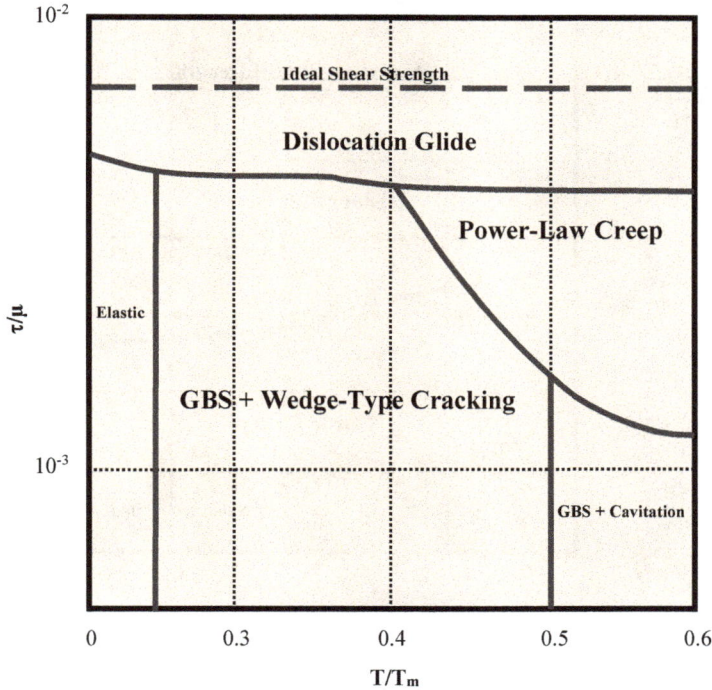

Figure 33. Deformation mechanism map for 1.23Cr-1.2Mo-0.26V steel

The establishment of a creep-fatigue life prediction method is important for the assessment of the remaining life of high-temperature components. Creep-fatigue tests with strain-hold durations which included 100 hours per cycle were carried out[36] on high-temperature component materials. The relationship between the creep-fatigue life and the damage evolution during strain-holds was examined. It was suggested that creep-damage during stress relaxation should be divided into matrix creep damage and grain-boundary creep damage, on the basis of a comparison of experimental results and the creep-deformation-mechanism map. A non-linear damage accumulation model was proposed which was based upon experimental results and theoretical considerations. The creep-fatigue lives of high-temperature materials could be predicted with high accuracy.

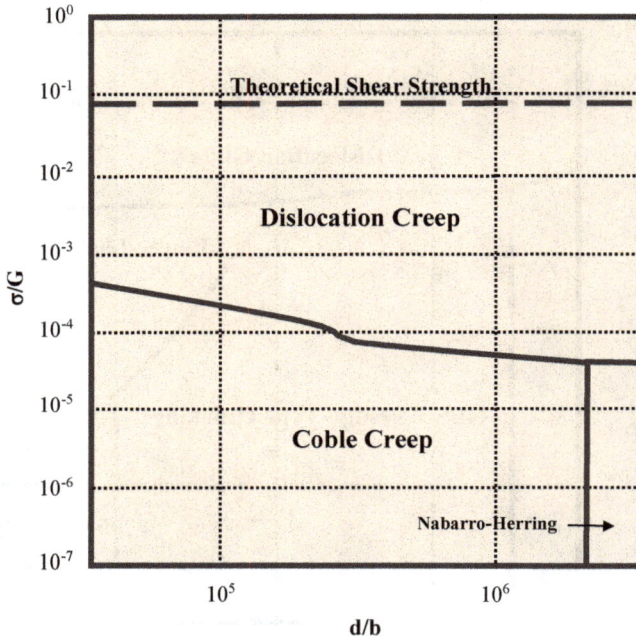

Figure 34. Deformation mechanism map for 25Cr-20wt%Ni austenitic stainless steel at an homologous temperature of 0.7

A critical review of previous creep studies concluded that traditional approaches such as steady-state behavior, power-law equations and the assumption that diffusional creep mechanisms were predominant at low stresses, should be seriously reconsidered[37]. In particular, creep strain-rate versus time curves showed that a decaying primary rate led on to an accelerating tertiary stage; thus giving a minimum rather than a secondary period. It was propose that conventional steady-state mechanisms should therefore be abandoned in favor of an understanding of the processes which govern strain-accumulation and of the damage phenomena which cause tertiary creep and fracture. It was pointed out that creep always takes place via dislocation processes, with no change to the diffusional creep mechanisms with decreasing stress; thus negating the concept of deformation-mechanism maps. Alternative descriptions were provided by normalizing the applied stress, using the ultimate tensile stress and yield stress at the creep temperature. In this way, the resulting Wilshire equations permitted an accurate prediction to be made of 100000h of creep data, by using property values which arose from tests lasting only 5000h, for a series of 2.25Cr steels.

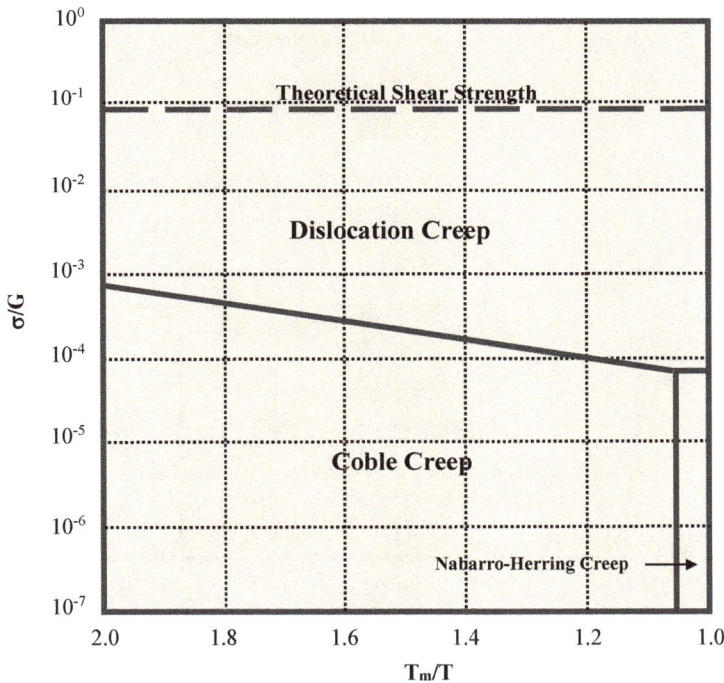

Figure 35. Deformation mechanism map for 25Cr-20wt%Ni austenitic stainless steel with a grain size of 40 μm

The domains of existence of deformation mechanisms in a map which was associated with phase transformation, and mechanical effects related to aging processes, were investigated[38] in austenitic stainless steels. The incidence of grain-boundary sliding, as both an additional deformation mechanism and a damage-accumulation process was considered. A predictive analysis of two typical high-temperature engineering systems was attempted on the basis of the map information. This prediction indicated the possibility of grain-boundary sliding and creep-strain jumps interfering with the expected operational life of components in systems operating at high temperatures.

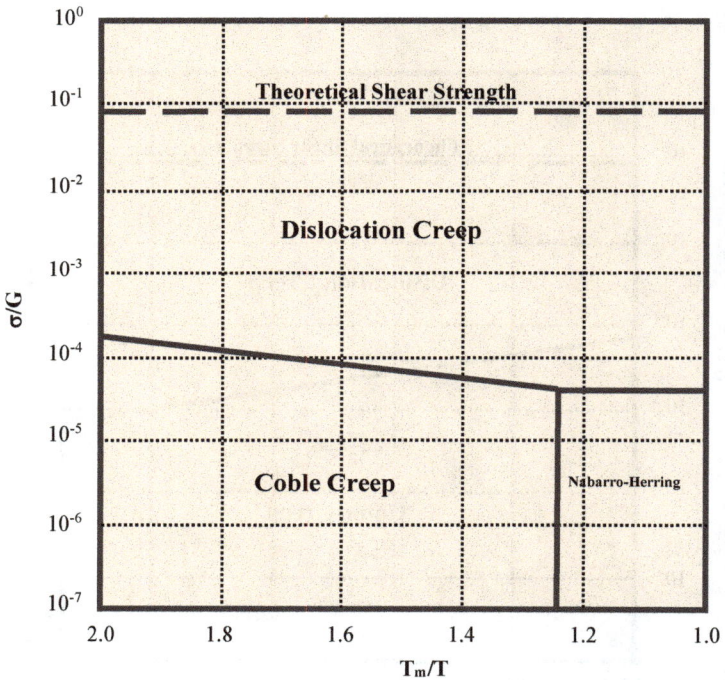

Figure 36. Deformation mechanism map for 25Cr-20wt%Ni
austenitic stainless steel with a grain size of 160µm

The deformation-mechanism map of 2.25Cr-1Mo steel was determined[39] by using creep data which had been obtained for a wide range of creep rates extending down to 10^{-1}/s.

The stress-dependence of the minimum creep-rates of the steel was similar to that of particle-strengthened materials: with high (H), intermediate (I) and low (L) stress regions.

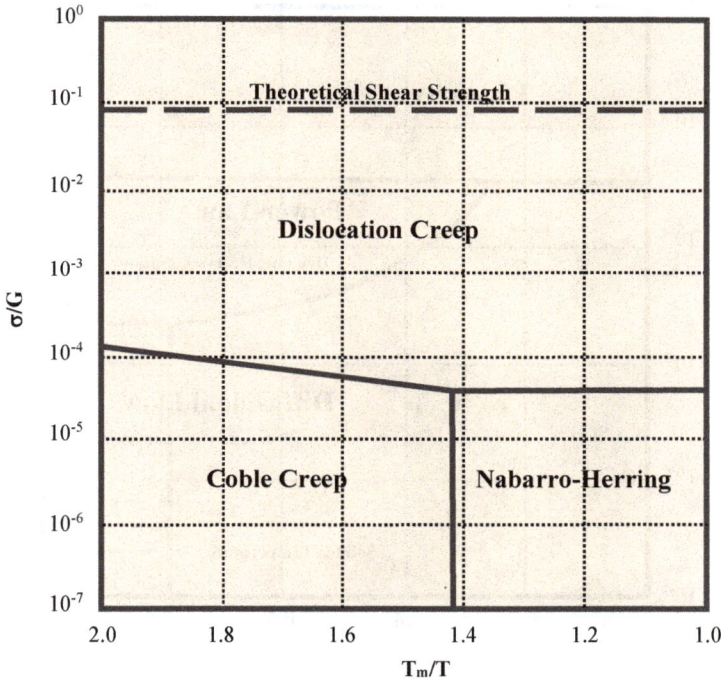

Figure 37. Deformation-mechanism map for 25Cr-20wt%Ni austenitic stainless steel with a grain size of 600μm

The stress-exponent and activation energy for the creep rate suggested that dislocation creep controlled by lattice diffusion was the deformation mechanism in regions I and L, which including typical service conditions for the steel. A transition to diffusion creep occurred at a lower creep-rate than that expected from the deformation-mechanism maps. Region H appeared above the athermal yield stress. During loading in this region, athermal plastic deformation took place via dislocation-glide, and then dislocation creep began. The dislocation creep in region H was different to that in regions I and L, due to the plastic deformation suffered during loading. A modified creep-mechanism map for 2.25Cr-1Mo steel was proposed on the basis of the experimental results.

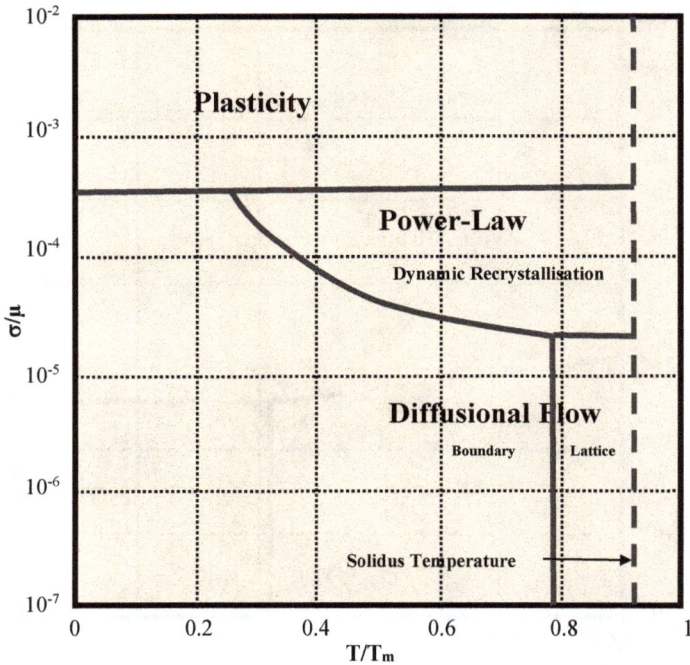

Figure 38. Deformation mechanism map for AISI304L

Deformation-mechanism maps for a 1CrMoV steel were used[40] to describe the damage which occurred during high-temperature cyclic-hold tests. A grain-boundary sliding regime was used to indicate the onset of wedge-crack formation. The strain-rate transitions which bounded regimes of known predominant damage mechanism were deduced for temperatures ranging from 450 to 650C. The transition strain-rate data were incorporated into a ductility-exhaustion approach to life-prediction. The transitions then permitted an accurate demarcation to be established between creep and fatigue damage. Cyclic life predictions which were based upon the ductility-exhaustion approach to cyclic-hold tests at 565C yielded excellent agreement with observations.

A deformation-mechanism map was constructed[41] for 1.23Cr-1.2Mo-0.26V rotor steel as a function of published temperature, stress and strain-rate creep-test results. In place of diffusional creep, grain-boundary sliding accommodated by various deformation processes predominated at lower strain-rates. The grain-boundary sliding-dominated region could be further sub-divided into two parts; one where grain-boundary sliding was accommodated by wedge-type cracking at homologous temperatures below 0.5, and one where the accommodation process changed to creep cavitation at homologous temperatures above 0.5. The form of the map was confirmed by experimental data and mathematical modelling. The latter could predict the predominance of various deformation mechanisms over wide ranges of stress and temperature.

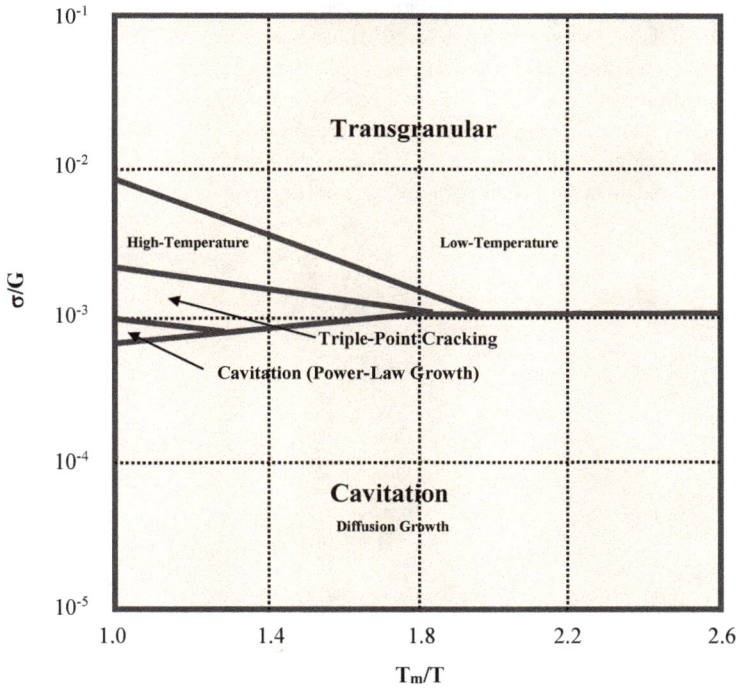

Figure 39. Deformation mechanism map for AISI316
stainless steel with a grain size of 150μm

Deformation-mechanism maps for a precipitate-free 25Cr-20wt%Ni austenitic stainless steel were expressed[42] in terms of normalized stress, reciprocal homologous temperature and normalized grain size. The maps were based largely upon experimental results rather than upon constitutive strain-rate equations predicted by deformation mechanisms. The maps systematically classified the complex creep behaviors of the stainless steel, clearly revealing the transitions in creep behavior with changes in stress, temperature and grain size. When such maps were prepared only on the basis of constitutive equations, they were not in agreement with experimental results.

The behavior of AISI304L stainless steel was studied[43] by using constant-stress tensile creep tests at 250 to 650C. No creep deformation was detected[44] between 250 and 400C. After an initial stress-dependent loading strain, no further deformation occurred. At 650C, creep curves having a characteristic shape were observed, with a relatively large loading strain being followed by a long primary-creep range. No steady-state creep behavior with a constant creep rate was observed. The creep-rate instead increased due to the superposition of primary and tertiary creep. Microstructural observations, performed following deformation to various points of the creep curve, showed that the development of sub-grains continued up to fracture[45]. An acceleration of the creep rate was caused by the formation and growth of cavities and wedge cracks.

Lead-

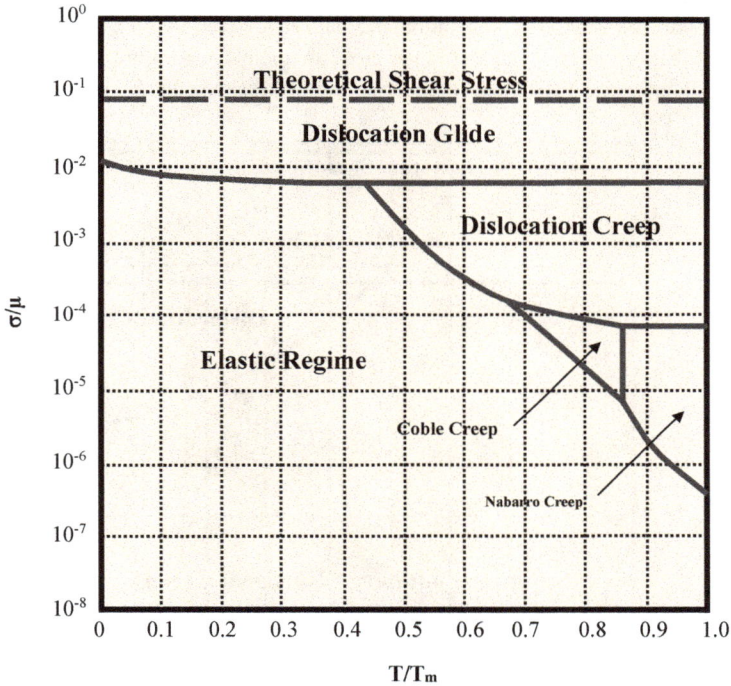

Figure 40. Deformation mechanism map for lead,
with a grain size of 32μm, at a critical strain rate of 10^{-8}/s

A deformation-mechanism map was constructed[46] for the mechanisms of strain relaxation in thin films of lead which had been deposited[47] onto oxidized silicon wafers at room temperature, and then thermally cycled between liquid helium and room temperatures. The stress level was deduced from strains measured using X-ray diffraction. During cooling, the strain relaxed rapidly in the field of dislocation glide for films which were thicker than 0.2μm. During heating, most of the strain was again probably relaxed by glide. In the case of a 0.5μm-thick film, the stress following primary relaxation was between 1 x 10^9 and 1.5 x 10^9dyne/cm^2 for cooling and between 0.17 x 10^9 and 0.24 x 10^9dyne/cm^2 for heating at 200 to 280K. Slow secondary relaxation occurred following primary relaxation. The compressive-strain relaxation-rate at room temperature was very close to the rate calculated on the basis of grain-boundary diffusion creep, thus suggesting that the secondary relaxation mechanism of compressive strain was grain-boundary diffusion creep at temperatures close to room temperature. Dislocation slip-lines were observed within grains, and hillocks were observed on grain boundaries.

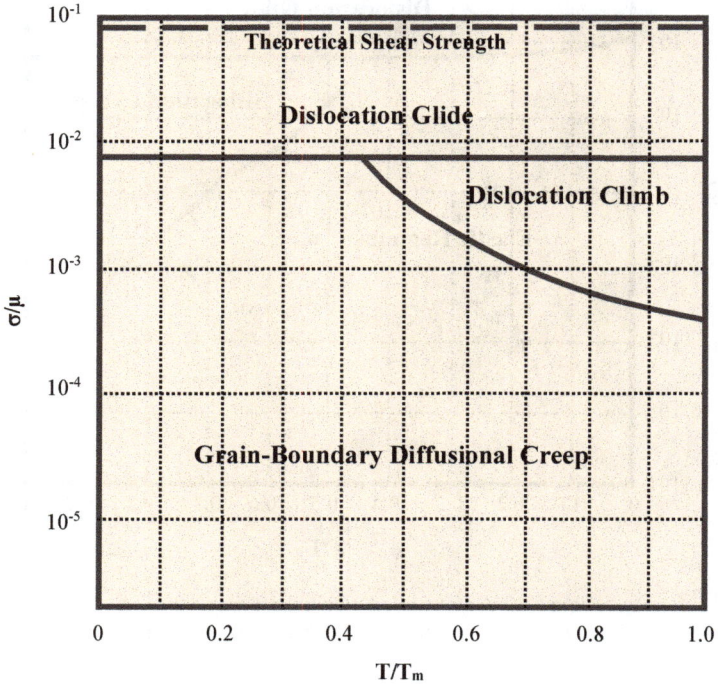

Figure 41. Calculated deformation mechanism map for lead thin film with a thickness of 0.5μm and a grain size of 2.5μm

The maximum-work principle was used[48] to analyze the axisymmetrical co-deformation in face-centered cubic crystals for twinning on (111)<112> and slip on {111}<112> systems. The influence of ξ, the ratio of critical resolved shear stress for twinning to slip, upon the yield stress states and corresponding active slip or/and twinning systems for orientations in the standard stereographic triangle of a cubic crystal was systematically investigated. The Taylor factors and the anisotropy of yield strength for three important orientations [100], [110] and [111] in orientation space were analyzed. It was found that the yield-strength asymmetry for the case of axisymmetrical deformation of tension and compression could be explained on the basis of the microscopic theory of crystal plasticity. The concept of an orientation factor for twinning ability was proposed, and the deformation-mechanism map in orientation space was established for the case of axisymmetrical deformation. The deformation-texture formation and the development of face-centered cubic crystals with low stacking-fault energy for axisymmetrical tension could be explained qualitatively on the basis of the results.

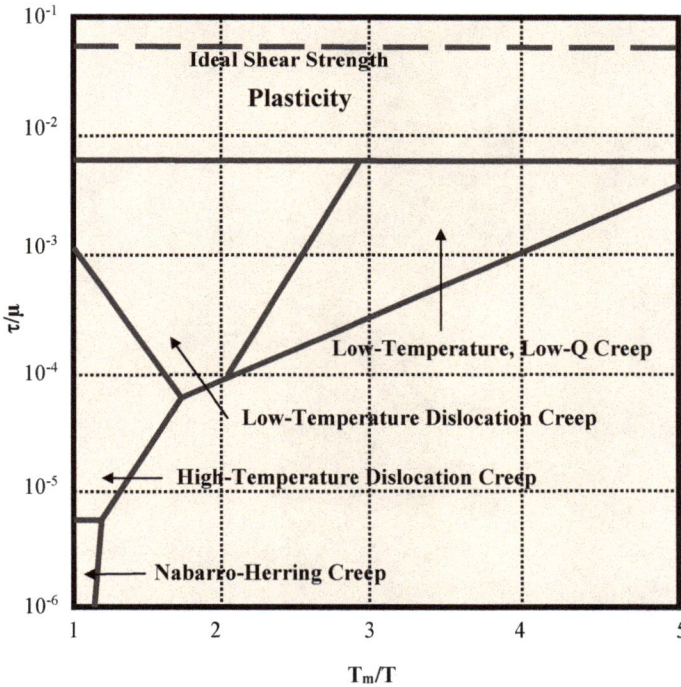

Figure 42. New deformation-mechanism map for 4N lead with a grain size of 310μm

Creep was investigated[49] in typical face-centered cubic metals at homologous temperatures below 0.3. All of the samples exhibited a marked creep behavior at these temperatures, with an apparent activation energy of 15 to 30kJ/mol and a stress exponent of 2 to 5. The results revealed a new creep region that had not appeared in other deformation-mechanism maps for pure face-centered cubic metals.

Magnesium-

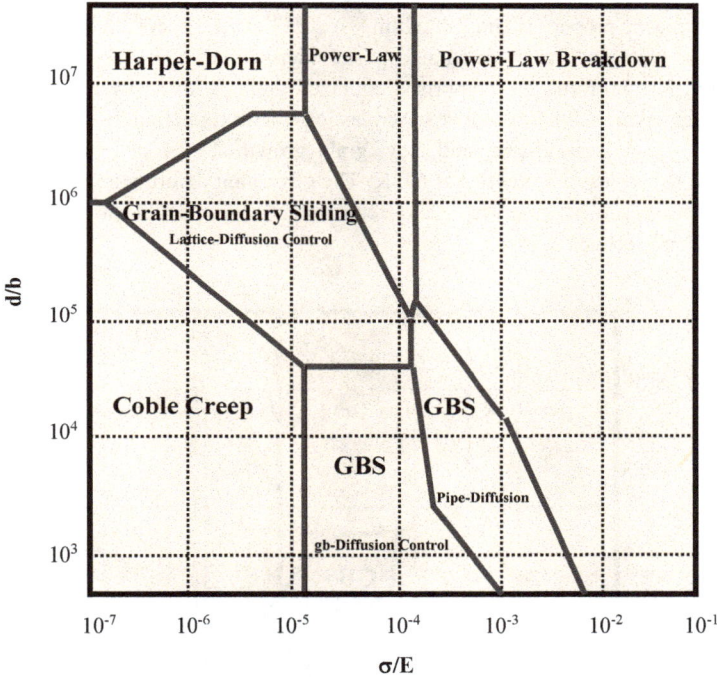

*Figure 43. Deformation mechanism map for pure
magnesium at an homologous temperature of 0.67*

The superplastic behavior of thin rolled sheets of AZ61 and AZ31 alloy was evaluated[50] at 573 to 693K, and was compared with as-received materials having relatively coarse grains. Existing deformation-mechanism maps for face-centered cubic and body-centered cubic metals were found to be deficient for the prediction of the deformation behavior of magnesium alloys having an hexagonal close-packed structure. By using experimental data for magnesium alloys which were associated with various competing deformation mechanisms at high temperatures, deformation-mechanism maps for such alloys could be constructed. Excellent predictive capabilities of the maps were demonstrated at 573 to 673K.

The superplastic properties and formability of AZ31 sheet, processed by strip-casting and subsequent warm-rolling, were studied[51]. The microstructure of sheet having a thickness of 1.3mm was uniform and comprised equiaxed grains with an average size of 6.6μm. Such sheets exhibited an excellent superplasticity, with a maximum elongation of 800% at 673K. A small grain-size, and slow grain growth during deformation, resulted in improved superplastic properties at 673K. The governing deformation mechanism which operated at a given strain-rate and temperature-range could be predicted by the deformation-mechanism maps.

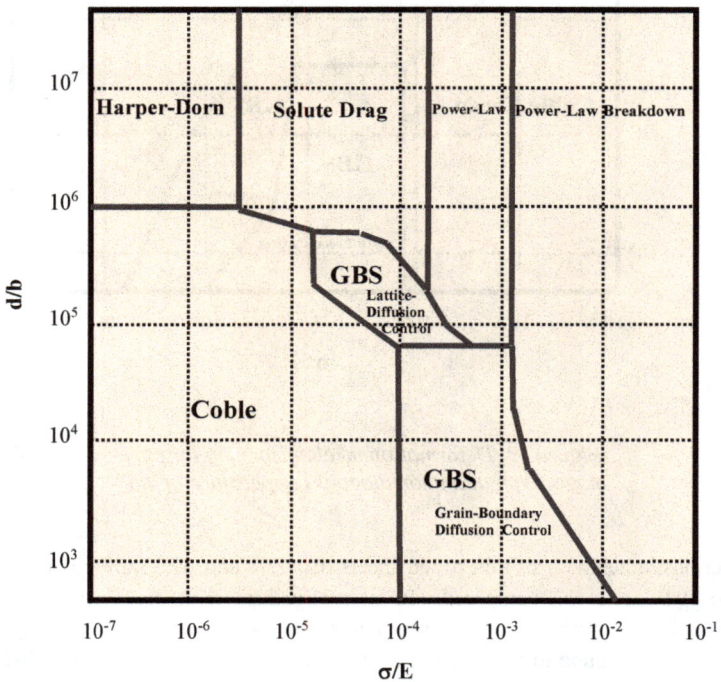

Figure 44. Deformation mechanism map for Mg–3Al–1%Zn at 573K

Materials Research Forum LLC
https://doi.org/10.21741/9781644901694

A study was made of the hot-deformation behavior of the as-cast Mg alloys, AM30 and AM50. The alloys were investigated[52] with regard to mechanical and microstructural evolution at 423 to 623K and strain-rates ranging from 10^{-3} to 10/s. A safe processing domain was identified for as-cast AM30 and AM50 alloys on the basis of the strain-rate sensitivity map. The deformed microstructures of both AM30 and AM50 revealed the occurrence of dynamic recovery and recrystallization at temperatures above 523K. Although the two alloys exhibited similar behaviors in the hot-deformation regime, they deviated at 623K and low strain-rate, where AM30 was fully recrystallized while AM50 exhibited a necklace structure following deformation. This was attributed to the difference in the rate of dynamic recovery in the two alloys, owing to the difference in aluminium concentration. Continuous dynamic recrystallization was the recrystallization mechanism which was active in the processing regime.

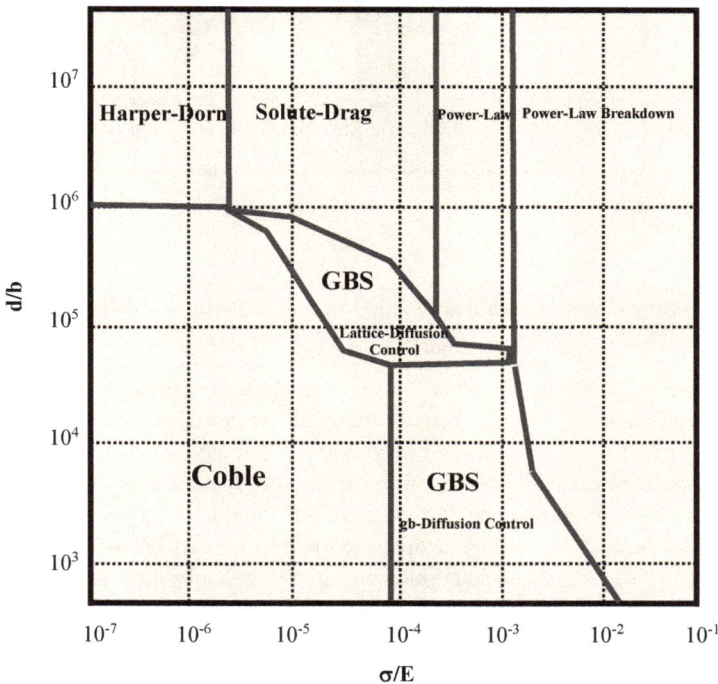

Figure 45. Deformation mechanism map for Mg–3Al–1%Zn at 623K

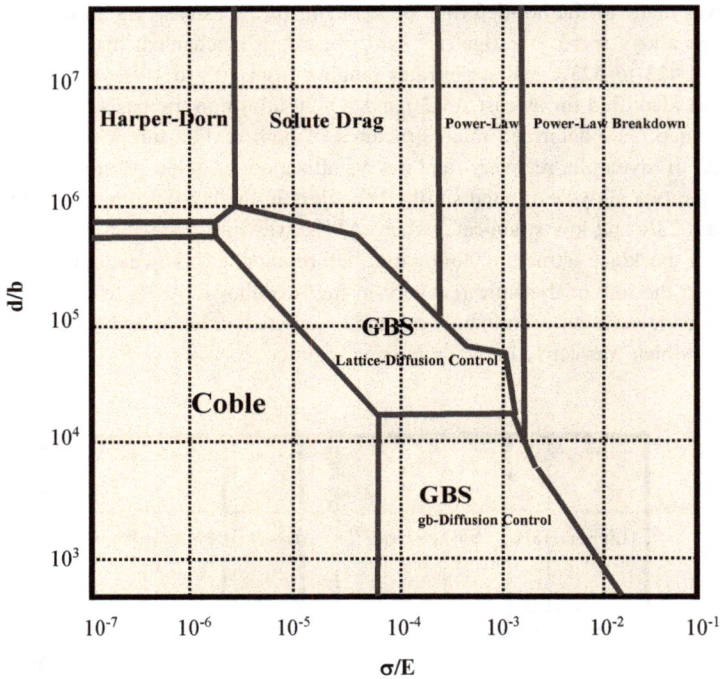

Figure 46. Deformation mechanism map for Mg–3Al–1%Zn at 673K

A comprehensive study[53] combined *in situ* scanning electron microscopy experiments and atomistic simulations in order to quantify the effect of crystal size upon the change in deformation modes of a-axis oriented magnesium single crystals at room temperature. The experimental results indicated that deformation was dominated by the nucleation and propagation of tensile twins. The stress which was required for twin propagation was found to increase with decreasing sample size, revealing a typical so-called smaller is stronger behavior. An anomalous increase in strain-hardening was reported, for microcrystals having diameters greater than ~18μm, which was induced by twin-twin and dislocation-twin interactions. The hardening rate gradually decreased toward the bulk response as the microcrystal size increased. Below 18 μm, the deformation was dominated by the nucleation and propagation of a single tensile twin, followed by basal-slip activity in the twinned crystal; leading to no apparent hardening. In addition, molecular dynamics simulations indicated a transition from twinning-mediated plasticity to dislocation-mediated plasticity for crystal sizes which were below a few hundred nanometres in size. A deformation-mechanism map for twin-oriented magnesium single crystals, ranging from the nano-scale to bulk, was proposed on the basis of the simulations and experiments. This predicted size-affected deformation mechanism for twin-oriented magnesium single crystals could lead to an improved understanding of the competition between dislocation plasticity and twinning plasticity.

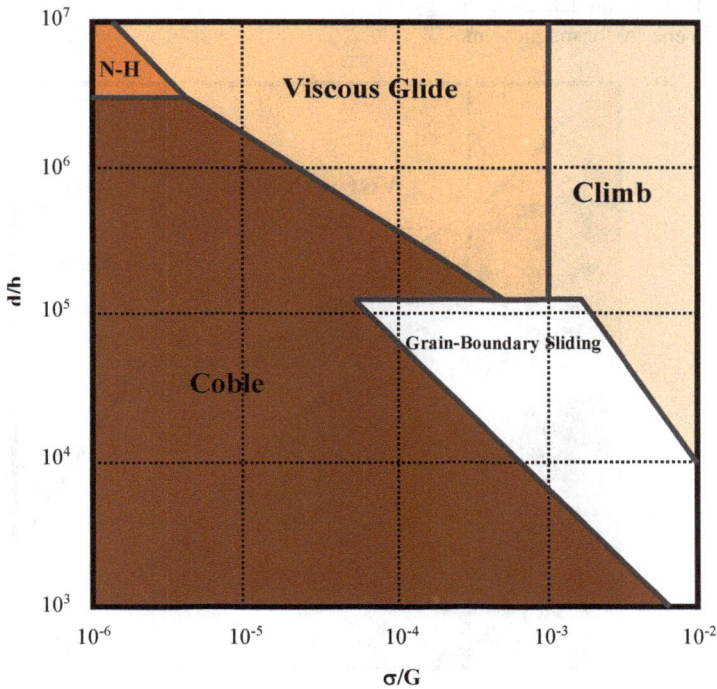

Figure 47. Deformation mechanism map for AZ31 magnesium alloy at 573K

Mg-6Li-3Zn alloy sheets were prepared[54] by melting and casting, and rolled to a total reduction of 94%. The high-temperature mechanical behavior, microstructures and deformation mechanisms were investigated. A maximum elongation to failure of 300%

was demonstrated at 623K and an initial strain-rate of 1.67 x 10^{-3}/s. Optical microscopy and transmission electron microscopy revealed that appreciable dynamic recrystallization and grain refinement occurred in banded grains at 573K and an initial strain-rate of 1.67 x 10^{-3}/s, where the sub-grain contour was ambiguous and the dislocation distribution was relatively uniform. It was shown by using the new deformation-mechanism map that the high-temperature deformation mechanism in Mg-6Li-3Zn alloy sheet with banded grains, at 573K and an initial strain-rate of 1.67 x 10^{-3}/s, was dislocation viscous glide controlled by lattice diffusion. The stress-exponent was 3 (strain-rate sensitivity exponent of 0.33) and the deformation activation energy was 134.8kJ/mol; the same as the lattice-diffusion activation energy for magnesium.

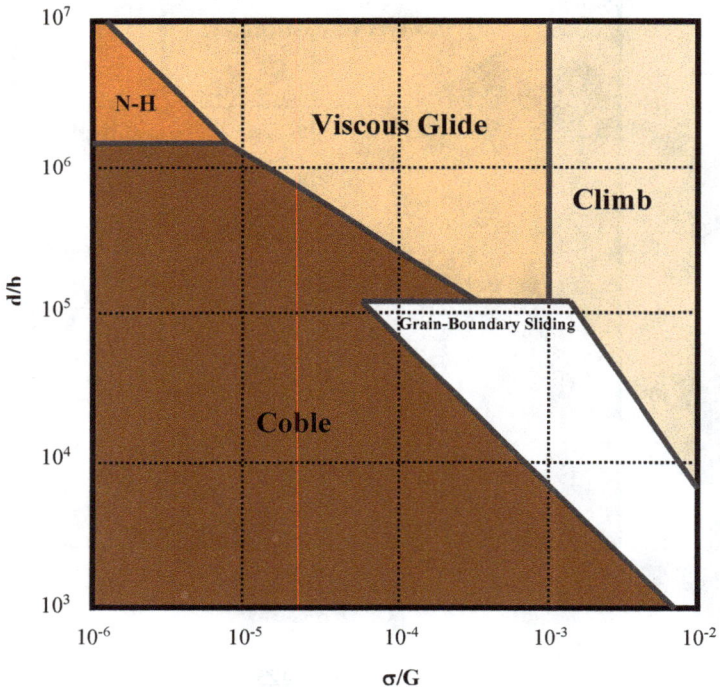

Figure 48. Deformation mechanism map for AZ31 magnesium alloy at 623K

Deformation-mechanism maps for 0 to 883K and shear strain-rates of 10^{-10} to 10^6/s were built up[55] from available rate equations for deformation mechanisms in pure magnesium or magnesium alloys. It was found that the grain size had little effect upon the fields of plasticity and phonon or electron drag, but it had a marked effect upon the fields of power-law creep, diffusion creep and Harper-Dorn creep for the above ranges of temperature, strain-rate and grain size. A larger grain size was predicted to be helpful in increasing the field range of power-law creep, but decreased that of diffusion-creep when the grain size was less than ~204 µm. Harper-Dorn creep dominated the deformation due to diffusion creep at grain sizes ranging from 204 to 255 µm.

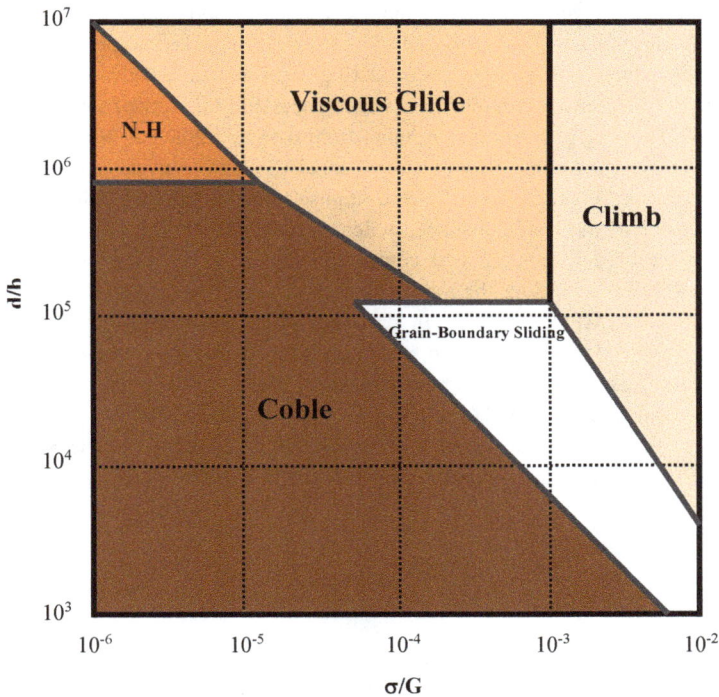

Figure 49. Deformation mechanism map for AZ31 magnesium alloy at 673K

The maps included only plasticity, phonon or electron drag and power-law creep when the grain size was greater than ~255μm, when the grain size had little influence upon the maps. Upon comparing the reported data for Mg-Gd-Y alloys with maps built up from available rate equations, it was concluded that the maps were an effective tool for predicting the deformation behavior of the Mg-Gd-Y alloys and for systematically classifying discrepancies in the deformation mechanism. Differences nevertheless existed in the deformation mechanisms of the alloys according to the reported data and that predicted by the maps. Refinement of the maps with regard to mechanical twinning and adiabatic shear was suggested to be necessary.

Stress-temperature curves at constant strain-rate were summarized[56] for the steady-state flow conditions of single crystals and polycrystals, and fields of predominating microscopic mechanisms were proposed. The superplastic deformation behavior of fine-grained AZ61 magnesium alloy sheet during equi-biaxial tensile deformation was investigated[57]. Thin circular diaphragms were successfully deformed into hemispherical domes at 673K, using gas pressures ranging from 0.46 to 1.20MPa. Within this pressure range, an average shell stress ranging from 7 to 23MPa and an average deformation-rate ranging from 2×10^{-4} to 5×10^{-3}/s were imposed on the deforming hemisphere. The thickness profile of the resultant shape, which was sensitive to the strain-rate sensitivity and the degree of deformation, was compared with an analytical model. Under low-pressure (0.46MPa) conditions the forming process obeyed the model, but this was not so at high (0.8 or 1.2MPa) pressures. In the latter cases, according to the deformation-mechanism map for magnesium, the rate-controlling deformation mechanism changed from lattice-diffusion controlled grain-boundary sliding to lattice-diffusion controlled dislocation-climb creep during forming; due to dynamic grain growth. The degree of uniformity of thickness distribution was less than that predicted by the theoretical model. The reduction in strain-rate sensitivity value resulted from this change in deformation mechanism.

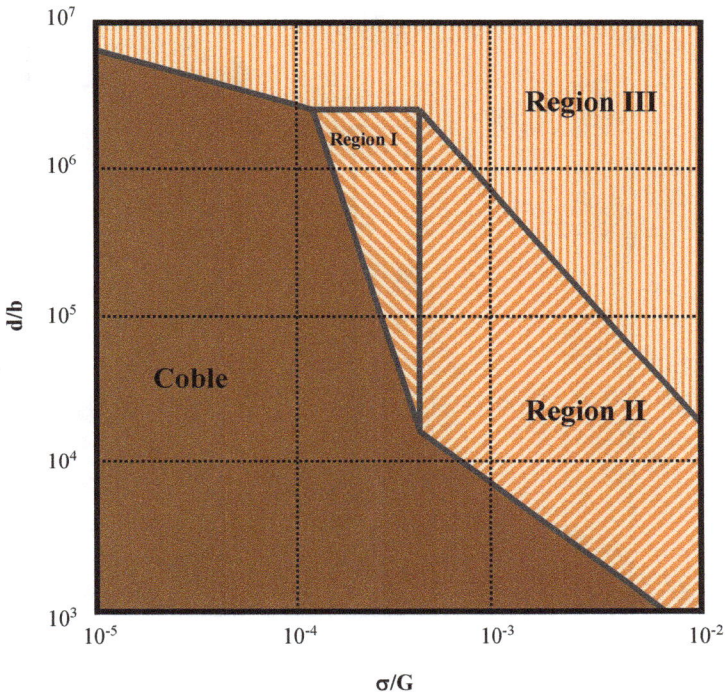

Figure 50. Deformation mechanism map of ZK60 magnesium alloy at 493K. Deformation in region I is controlled by a stress exponent of 5, deformation in region II is controlled by a stress exponent of 7 and deformation in region III is again controlled by a stress exponent of 5

The high-temperature deformation behavior of AZ91 alloy with ultrafine grains was studied[58] at 473 to 573K. Analyses based upon the deformation mechanism maps for magnesium, and a comparison of the experimental data with the values predicted by equations for several deformation mechanisms, indicated that Coble creep was the rate-controlling deformation mechanism in the low strain-rate range. The critical grain size at which Coble creep could be detected in magnesium at a given strain-rate and temperature was calculated.

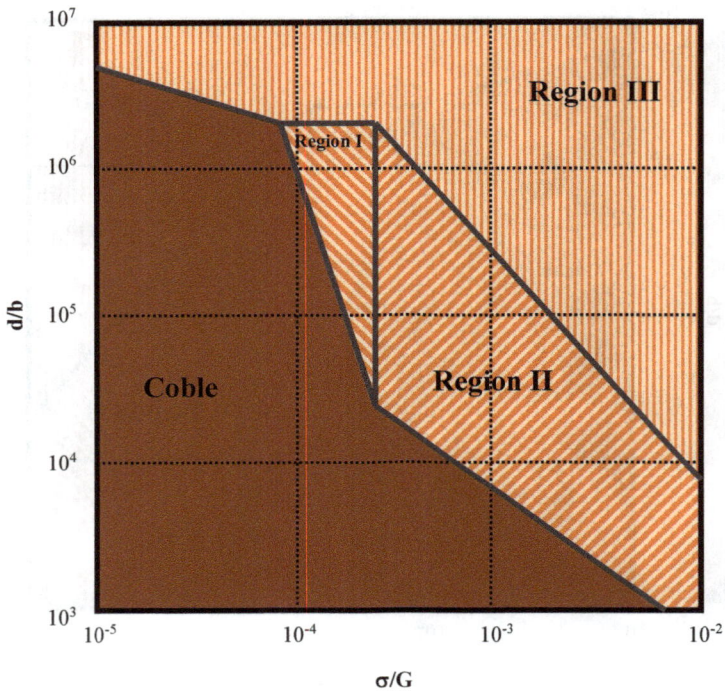

Figure 51. Deformation mechanism map of ZK60 magnesium alloy at 573K. Deformation in region I is controlled by a stress exponent of 5, deformation in region II is controlled by a stress exponent of 7 and deformation in region III is again controlled by a stress exponent of 5

Fine-grained sheets of Mg-7.83%Li and Mg-8.42%Li alloys were prepared by melting and casting, heavy rolling (reduction of more than 92%) and nitrate-bath annealing, and their superplasticity, microstructure, cavitation, fracture morphology and deformation mechanism were investigated[59]. The diffusivities and Gibbs free energy of the α-phase (5.7%Li) and β-phase (11%Li) at 573K were calculated in order to deduce the reason for superplastic grain growth. The results showed that a few cavities were distributed randomly and in isolation along the gauge length of Mg-8.42Li alloy at 573K and 1.67 x 10^{-3}/s. Transgranular fracture appeared at 573K and 5 x 10^{-4}/s in Mg-8.42Li alloy and dimple fracture along the grain boundary appeared at 573/K and 1.67 x 10^{-3}/s in Mg-

7.83Li alloy. A maximum superplasticity of 850% and 920% was obtained in Mg-7.83%Li and Mg-8.42%Li, respectively. Obvious superplastic grain growth at 573K appeared in Mg-7.83Li alloy. A comparison of normalized experimental data with a deformation-mechanism map which incorporated dislocation densities within grains revealed that the dominant deformation mechanisms in the two alloys were grain-boundary sliding, accommodated by slip that was controlled by lattice diffusion.

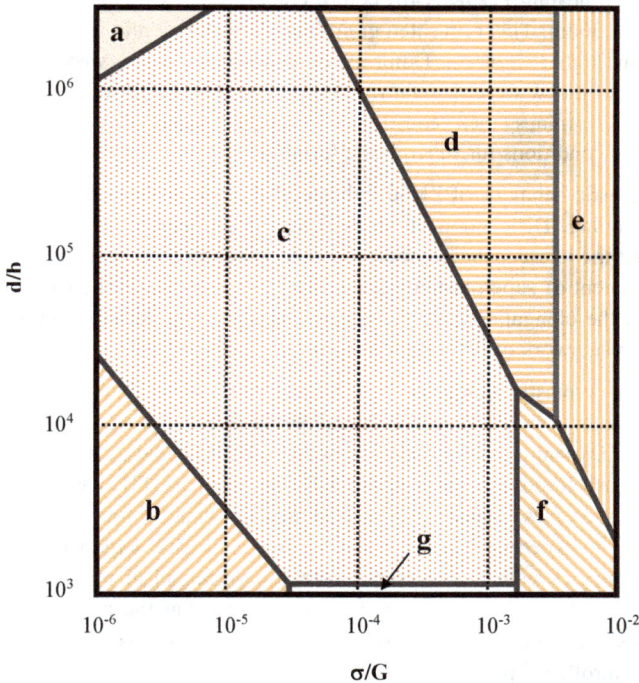

Figure 52. Deformation mechanism map for Mg–Li at 573K; a – Harper-Dorn slip controlled by lattice-diffusion and a stress exponent of 1, b – diffusion creep controlled by grain-boundary diffusion and a stress exponent of 1, c – superplastic grain-boundary sliding controlled by lattice-diffusion and a stress exponent of 2, d – dislocation slip controlled by lattice-diffusion and a stress exponent of 5, e - dislocation slip controlled by lattice-diffusion and a stress exponent of 7, f – dislocation pipe grain-boundary sliding controlled by pipe-diffusion and a stress exponent of 4, g – superplastic grain-boundary sliding controlled by grain-boundary diffusion and a stress exponent of 2

Materials Research Forum LLC
https://doi.org/10.21741/9781644901694

Double-shear creep testing was used[60] to study the creep behavior of AZ31 alloy which had been processed by equal-channel angular pressing in order to produce an average grain size of about 2.7μm. Rapid creep rates were observed in the early stages of deformation, due to grain-boundary sliding in the fine-grained structure. The creep rate decreased with increasing deformation, due to grain growth. The stress exponent for flow in the early stages was about 2 and the activation energy was some 92kJ/mol. These values were consistent with those expected for grain-boundary sliding under superplastic conditions. Annealing (723K, 24h) before creep-testing led to a markedly greater grain size of about 50μm. This prevented grain-boundary sliding and led to an increased stress exponent at higher stresses. Deformation-mechanism maps were constructed which incorporated the new experimental results for fine-grained alloy, plus published data for the alloy. The maps then constituted a useful means for choosing the required experimental conditions for superplastic forming.

The superplastic behavior of ZK60 indicated[61] the occurrence of deformation via grain-boundary sliding. Dislocation climb became the rate-controlling mechanism at higher stresses, but the rate-controlling mechanism at lower stresses remained unclear. Study of the development of superplasticity showed that an increase in the stress exponent and a decrease in the elongation occurred at low stresses. The deformation-mechanism maps exhibited a patchwork of deformation regimes.

Maximum superplastic elongations of 850 and 920% were observed[62] in Mg-7.83Li and Mg-8.42Li alloys at 573K, using initial strain-rates of 1.67 x 10^{-3} and 5 x 10^{-4}/s, respectively. Dynamic grain growth appeared at 573K. The differing atomic mobilities of magnesium and lithium in the α-phase 5.7Li and β-phase 11Li, and the differing Gibbs free energies of these phases contributed to the dynamic grain growth at 573K. By comparing the experimental stress exponents, grain-size exponents and deformation activation energies with constructed deformation-mechanism maps which incorporated dislocation processes within grains, it was deduced that the predominant deformation mechanisms in the alloys were grain-boundary sliding that was accommodated by lattice-diffusion controlled slip.

Nickel-

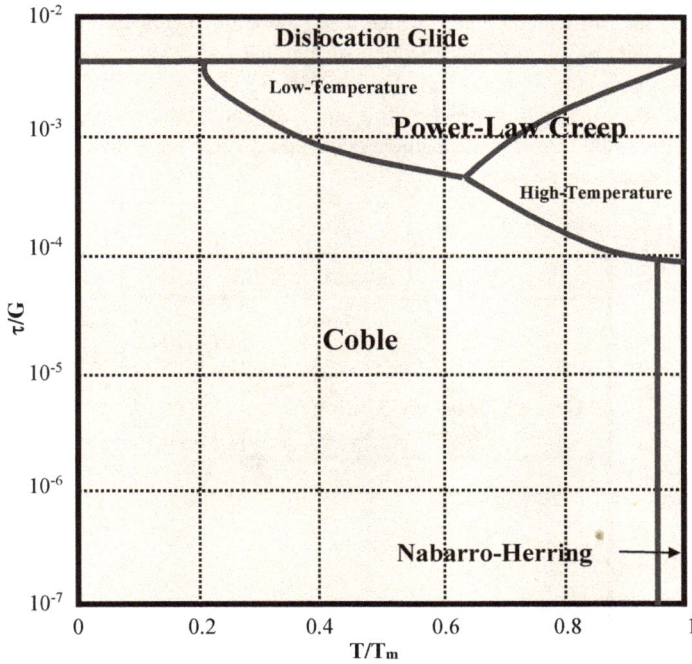

*Figure 53. Deformation mechanism map for nickel
with a grain size of 10μm as originally conceived*

Tensile tests over a wide range of strain rates (1.04 x 10^{-6} to 1.04/s) were performed[63] on electrodeposited nanocrystalline nickel and Ni-Co alloys with various grain sizes in order to investigate systematically the coupling effects of strain rate and grain size upon mechanical behavior. It was found that the grain size significantly affected the dependence of mechanical responses upon the imposed strain rate. In particular, peculiar fluctuations of flow stress and elongation-to-failure with changes in strain-rate were observed in the Ni-8.6wt%Co alloy with the finest grain size. By carefully analyzing the changes in strain-rate sensitivity exponent and apparent activation volume, such unique phenomena were rationalized in terms of a synergy of grain size and strain-rate with regard to the transition in deformation mechanisms including thermally-activated and mechanically-driven grain boundary activities, interactions of dislocations with grain boundaries and diffusional creep; which acted alone or in concert so as to dominate the plasticity. Building upon a classification of the characteristics of plastic deformation at the nanoscale, a two-dimensional deformation-mechanism map was proposed which elucidated the interactive effects of grain size and strain-rate on the deformation mechanisms and the related mechanical behaviors of face-centered-cubic nanocrystalline nickel and nickel-based alloys.

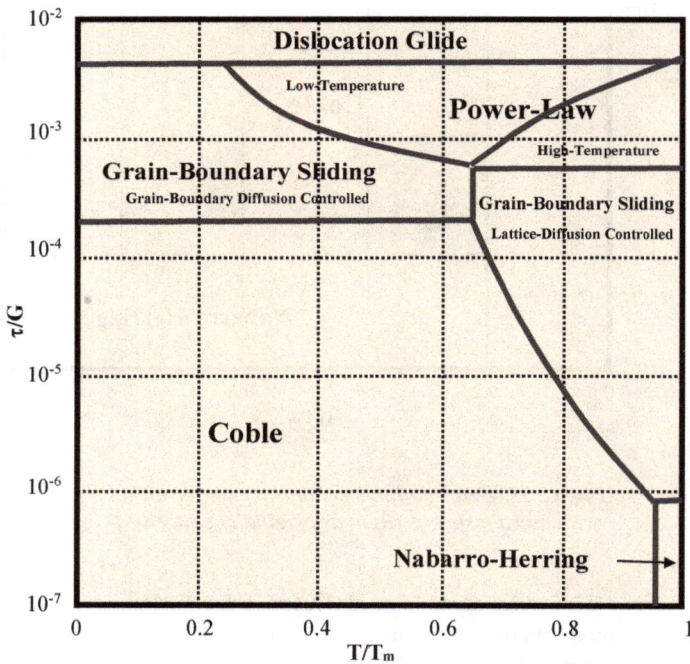

Figure 54. Deformation mechanism map for nickel with a grain size of 10μm, with the inclusion of grain-boundary sliding

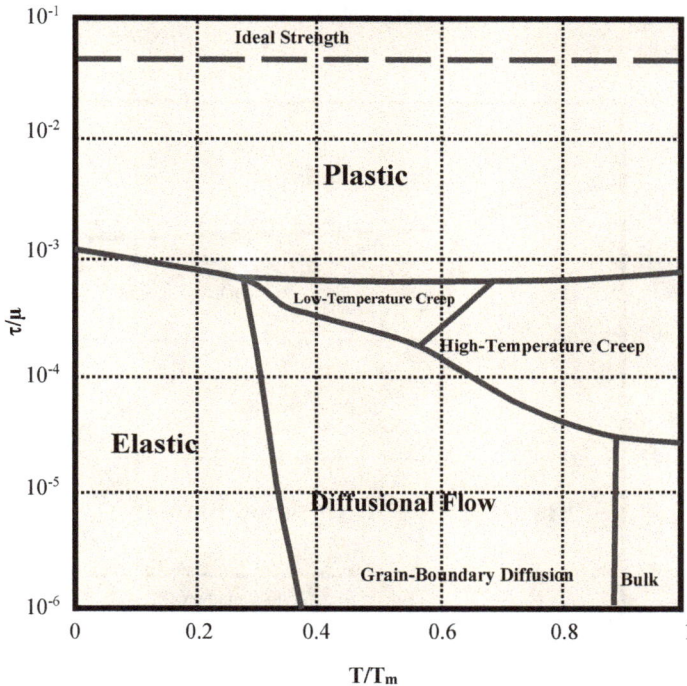

Figure 55. Deformation mechanism map for solid nickel with a grain size of 30μm

The results of high-temperature grain-boundary engineering of an experimental low stacking-fault energy nickel-based superalloy were compared[64] with those for a commercially available superalloy, RR1000. Deformation-mechanism maps for thermomechanical processing were compared along with the resultant length-fractions of Σ3 boundaries following sub-solvus and super-solvus annealing. Compared with the hot-deformation processing characteristics of RR1000, lowering the stacking-fault energy reduced the dislocation mobility and expanded the range of temperatures and strain-rates over which dislocation-based plastic flow mechanisms were operative in the low stacking-fault alloy. For both alloys, processing conditions which were conducive to dislocation-based plasticity allowed for the storage of strain-energy, within the microstructure, that was utilized for strain-induced boundary migration and the formation of Σ3 boundaries upon annealing. Based upon the results, alloying changes that serve to reduce the stacking-fault energy of nickel-based superalloys also make the alloys more amenable to grain-boundary engineering techniques that involve hot-deformation.

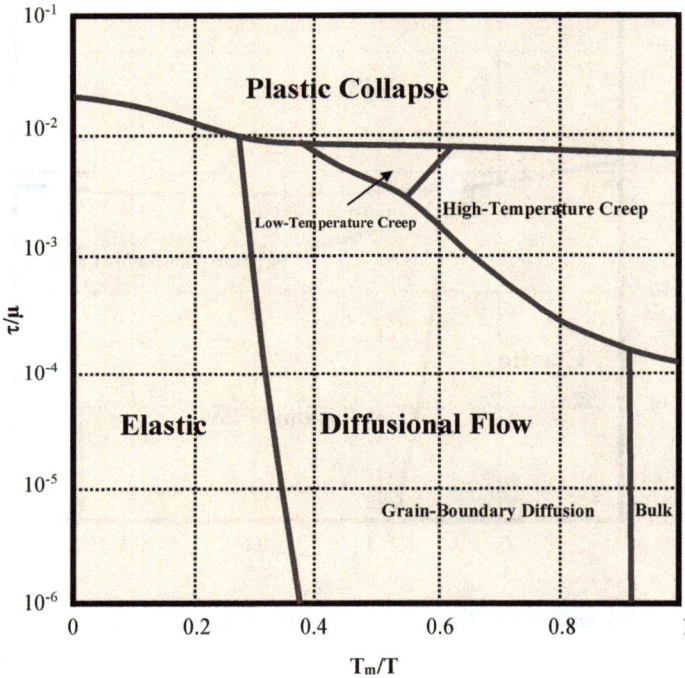

Figure 56. Deformation mechanism map of nickel foam with a relative density of 20%, a cell-size of 400μm and a grain size of 20μm

The creep-deformation mechanisms at intermediate temperature in ME3, a modern nickel-based disk superalloy, were investigated[65] using diffraction contrast imaging. Both conventional transmission electron microscopy and scanning transmission electron microscopy were utilized. Distinctly different deformation mechanisms became operative during creep at temperatures between 677 and 815C and at stresses ranging from 274 to 724MPa. Both polycrystalline and single-crystal creep tests were conducted. The single-crystal tests provided new insight into grain orientation effects upon the creep response and deformation mechanisms. Creep at temperatures of up to 760C resulted in thermally-activated shearing modes such as micro-twinning, stacking-fault ribbons and isolated superlattice extrinsic stacking-faults. On the other hand, these faulting modes occurred much less frequently during creep at 815C under lower applied stresses. The principal deformation-mode was instead dislocation-climb by-pass. In addition to the difference in creep behavior and creep-deformation mechanisms as a function of stress and temperature, it was also observed that microstructural evolution occurred during creep at 760C and above, where the secondary γ'-phase coarsened and the tertiary γ'-phase precipitates dissolved. Based upon this work, a creep-deformation mechanism-map was proposed which emphasized the influence of stress and temperature upon the underlying creep mechanisms.

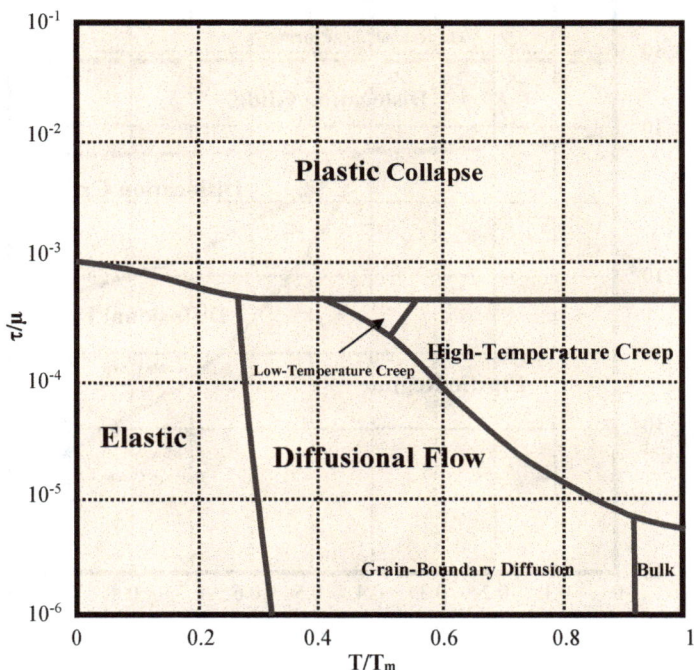

Figure 57. Deformation mechanism map for nickel foam with a relative density of 4%, a cell-size of 400μm and a grain size of 20μm

The effects of thermo-mechanical processing parameters upon the resulting microstructure of an experimental nickel-based superalloy containing 24wt%Co were investigated[66]. Hot-compression tests were performed at temperatures ranging from 1020 to 1100C and strain-rates ranging from 0.0005 to 0.1/s. The mechanically deformed samples were also subjected to annealing treatments at sub-solvus (1115C) and super-solvus (1140C) temperatures. The aim was to quantify and understand the behavior and evolution of both the grain-boundary structure and length-fraction of $\Sigma 3$ twin boundaries in the low stacking-fault energy superalloy. Over the above range of deformation parameters, the corresponding deformation-mechanism map revealed that dynamic recrystallization or dynamic recovery was dominant.

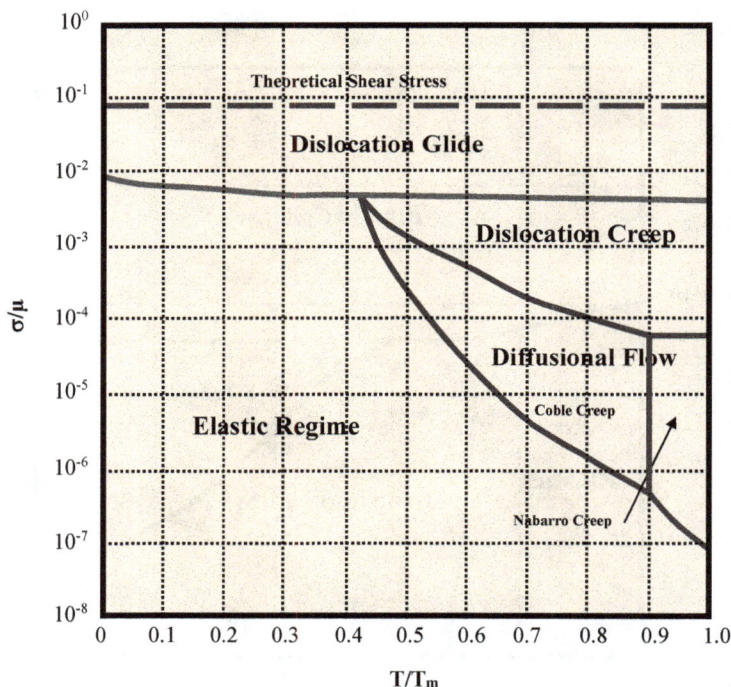

Figure 58. Deformation mechanism map for nickel,
with a grain size of 32 µm, at a critical strain rate of $10^{-8}/s$

These conditions largely promoted post-deformation grain refinement and the formation of annealing twins following annealing. Samples which were deformed at strain-rates of 0.0005 and 0.001/s at 1060 and 1100C exhibited extensive grain-boundary sliding/rotation that was associated with superplastic flow. Upon annealing, deformation conditions that resulted predominately in superplastic flow were found to provide negligible enhancement of twin boundaries and produced little or no post-deformation grain refinement.

A deformation mechanism map for a nickel-based superalloy was presented[67] for cyclic loading at low (300C), intermediate (550C) and high (700C) temperatures and low (0.7%) and high (1.0%) applied strain amplitudes. Strain-mapping was performed using digital image correlation during interrupted fatigue experiments at elevated temperatures and 1, 10, 100 and 1000 cycles, for each specified loading and temperature condition. The digital image correlation measurements were performed in a scanning electron microscope, which permitted high-resolution measurements to be made of heterogeneous slip events and a vacuum environment ensured stability of the speckle pattern for digital image correlation at high temperatures. The cumulative fatigue experiments showed that the slip bands were present in the first cycle and intensified with increasing number of cycles; resulting in highly localized strain accumulation. The strain mapping results were combined with microstructure characterization via electron back-scatter diffraction. The combination of crystal orientations and high-resolution strain measurements was used to determine the active slip planes. At low temperatures, slip bands followed the {111} octahedral planes. As the temperature increased however both the {111} octahedral and {100} cubic slip planes accommodated strain. The activation of cubic slip via cross-slip within the ordered intermetallic γ'-phase was well-documented in nickel-based superalloys and is generally accepted as being the mechanism responsible for the anomalous yield phenomenon. The present results represented an important quantifiable study of cubic slip-system activity at the mesoscale in polycrystalline γ-γ' nickel-based superalloys and was a key advance in calibrating the thermal activation components of polycrystalline deformation models.

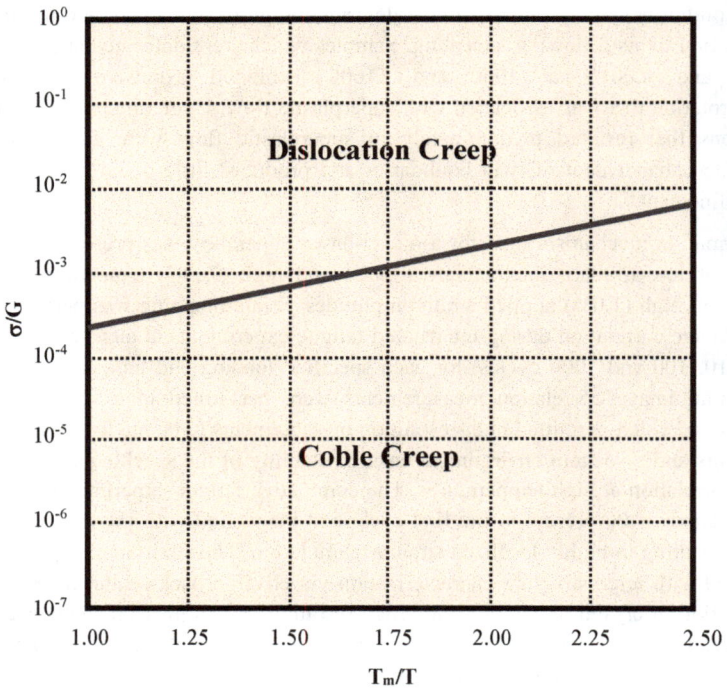

Figure 59. Deformation mechanism map for NiAl with a grain-size of 0.34µm

Material design and property modifications are traditionally associated with compositional changes, but subtle changes in the manufacturing process parameters can also have a dramatic effect upon the resultant material properties. An integrated computational materials engineering framework[68] was adopted in order to tailor the fatigue performance of the nickel-based RR1000. An existing fatigue model was used to identify those microstructural features which promote an enhanced fatigue life: that is, a uniform fine grain-size distribution, a random orientation, a distinct grain-boundary distribution (especially a high twin-boundary density and limited low-angle grain boundaries).

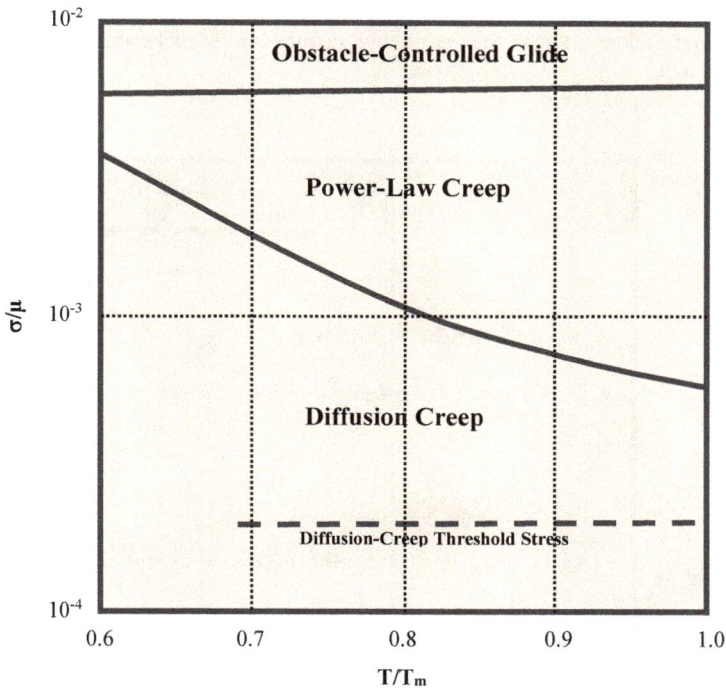

Figure 60. Deformation mechanism map for IN738LC
with an average grain size of 0.3mm

A deformation mechanism map and process models for the grain-boundary engineering of RR1000 were used to identify the optimum thermo-mechanical processing parameters for the obtention of desirable microstructural features. Small-scale forgings of RR1000 were produced and heat-treated so as to produce fine-grained or coarse-grained microstructures that reproduced conventionally processed and grain-boundary engineered conditions, respectively. For each of the four microstructural variants of RR1000, the twin density and grain size were characterized and were in agreement with the desired microstructural attributes. In order to validate the deformation mechanisms and fatigue behavior of the material, high-resolution digital image correlation was performed in order to generate strain-maps for the microstructural features. A high density of twin boundaries was confirmed to inhibit the length of slip bands, and this was directly linked to an extended fatigue life. This demonstrated the successful role of models, for both process and performance, in the design and manufacture of nickel-based superalloy disk forgings.

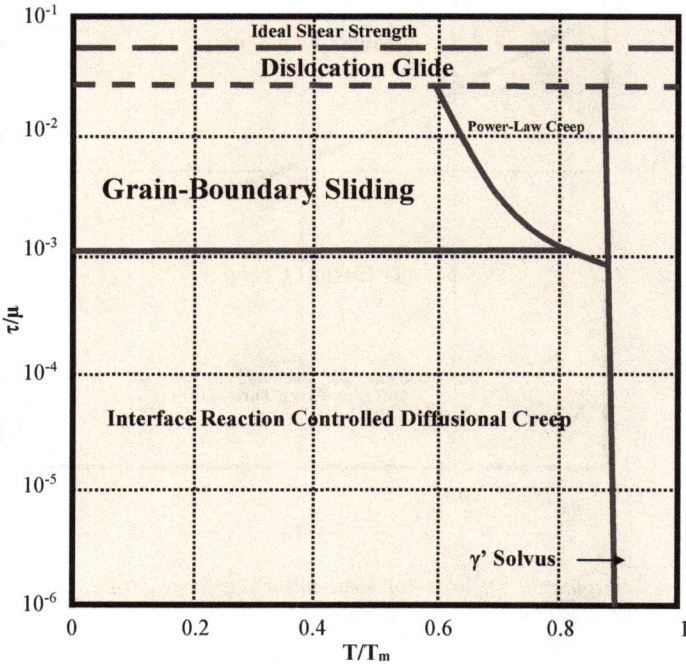

Figure 61. Deformation mechanism map for IN738LC turbine blades

The critical rotating components used in the hot section of gas turbines are subjected to cyclic loading conditions during operation, and the life of these structures is governed by their ability to resist fatigue. Since it is well known that microstructural parameters such as the grain size can significantly influence the fatigue behavior of the material, the conventional processes which are involved in the manufacture of these structures are carefully controlled in an effort to engineer the resultant microstructure. Again for the nickel-based RR1000, the development of process models and deformation mechanism maps has permitted not only control of the resultant grain size but also the ability to tailor and manipulate the resultant grain-boundary character-distribution. The increased level of microstructural control was coupled with a physics-based fatigue model in order to form an integrated computational materials engineering framework that was then used to guide the design of damage-tolerant microstructures. Simulations of a three-dimensional crystal-plasticity finite-element model were used to identify the microstructural features which are associated with strain-localization during cyclic loading and to guide the design of polycrystalline microstructures which are optimized for fatigue resistance.

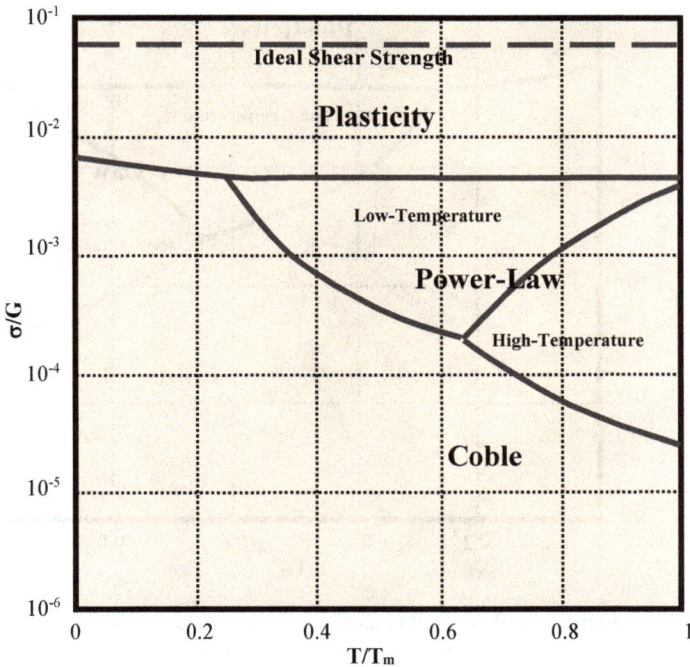

Figure 62. Deformation mechanism map for Incoloy 800H with an average grain size of 100μm

Conventionally processed and grain-boundary engineered forgings of RR1000 were again produced in order to validate the design methodology. For nominally equivalent grain sizes, high-resolution strain-maps, generated via digital image correlation, confirmed that a high density of twin boundaries in the grain-boundary engineered material were desirable microstructural features as they contributed to limiting the overall length of the persistent slip bands that often serve as precursors for the nucleation of fatigue cracks.

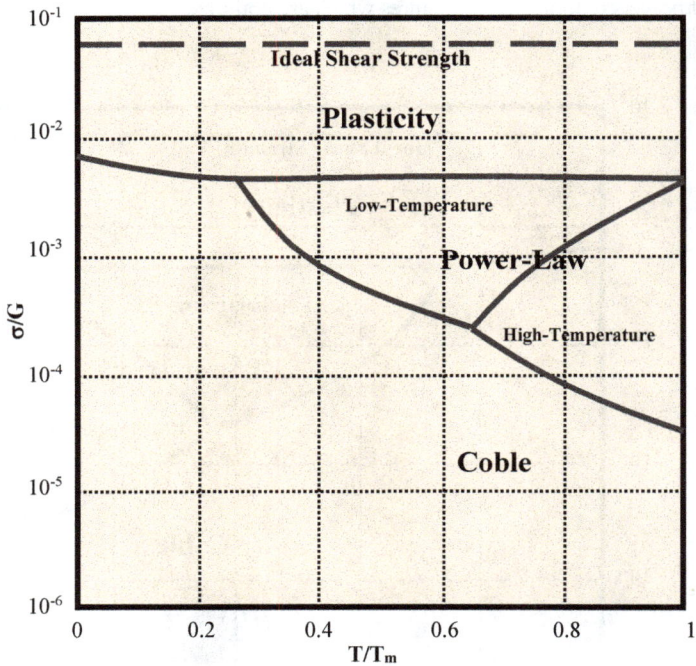

*Figure 63. Deformation mechanism map for Incoloy 800H
with an average grain size of 50 μm*

A study was made of the low-cycle fatigue behavior of a new nickel-based superalloy in the temperature range of 650 to 760C[69]. Two initial microstructures having differing precipitate distributions were produced, and the low-cycle fatigue behavior of these microstructures was characterized by using mechanical parameters such as cyclic anisotropy, plastic strain accumulation and back-stress evolution. The deformation response was confirmed via detailed electron microscopy. A deformation-mechanism map which illustrated the defect activity occurring under various testing conditions was constructed on the basis of the microscopy results.

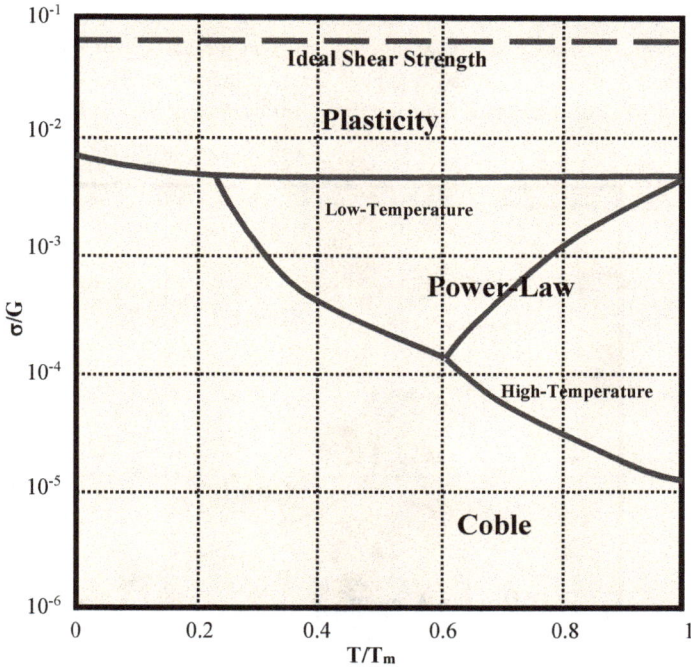

Figure 64. Deformation mechanism map for Incoloy 800H with an average grain size of 350μm

Amorphous Ni–Mo–P films were successfully deposited onto CoCrFeNi high-entropy alloys by electroless plating with the aim[70] of reviewing the ambient-temperature mechanical properties, plastic deformation and fracture mode of the coating|substrate system. The yield strength and ultimate tensile strength of a specimen which had been plated for 1h were 260 and 760MPa, respectively, with a 44.4% (yield strength) and 14.2% (ultimate tensile strength) improvement as compared with the uncoated high-entropy alloy's yield strength and ultimate tensile strength of 180MPa and 560MPa, respectively. This was attributed to the back-stress and repulsive image stress having suppressed the nucleation and mobility of dislocations. Due to substrate confinement, the deformation mode gradually changed from cracking (with homogeneous flow) to shear banding (with homogeneous flow). Fully-homogeneous flows eventually occurred, with decreasing film thickness, and a deformation-mechanism map was developed so as to express the transition in deformation mode. Brittle fracture at a high velocity, originating from the amorphous films, led to cleavage-cracking on the surface of the high-entropy alloy substrate. A thicker film had a higher crack velocity and caused deeper crack penetration into the substrate, thus reducing the substrate ductility.

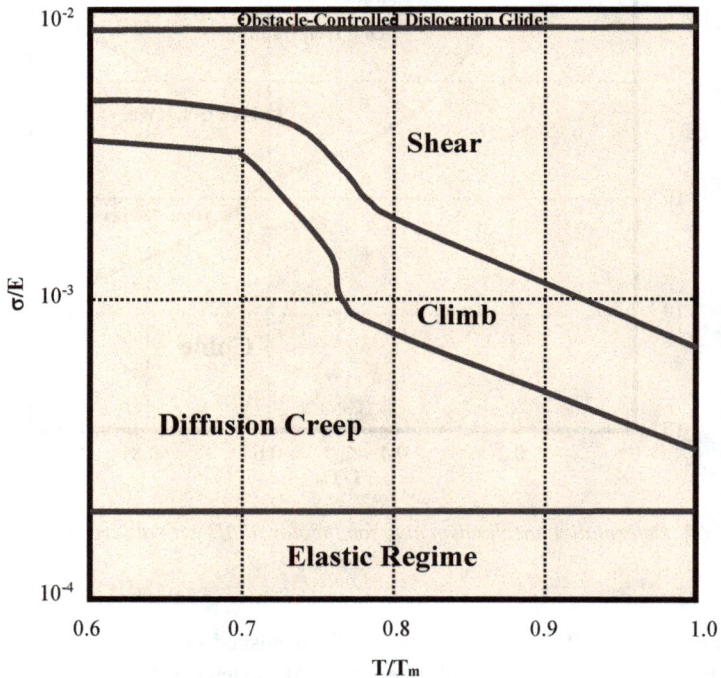

Figure 65. Deformation mechanism map for GTD-111 with a grain size of 1.2mm

A novel nickel-based superalloy was designed which had γ'-phase precipitate-strengthening, controlled γ/γ' lattice-misfit and a high configurational entropy of the γ-phase matrix, thus imparting an improved high temperature performance up to 800C[71]. The alloy contained nano-sized γ'-phase precipitates, MC and other grain-boundary carbides. Three variants of the alloy were fabricated using vacuum induction melting, a computationally-based homogenization cycle for reducing solidification segregation and the thermo-mechanical processing steps of forging followed by hot-rolling to produce plates from which standard test specimens were extracted. Tensile testing at room temperature, and at temperatures of up to 800C, revealed a yield stress which was superior to that of Nimonic 105. The three variants of the alloy were machinable, with maximum stresses that were comparable to standard results, as indicated by deformation-mechanisms maps deduced using Gleeble testing and electron back-scatter scatter diffraction. Due to the composition of experimental alloys falling outside of the typical range used to tabulate thermodynamic data, differences in phase predictions and related temperatures were detected between experimental and Thermo-Calc predictions. The γ'-phase forming elements, titanium and niobium, had a similar effect upon the γ'-phase precipitates and indirectly contributed to a change in the entropy of the γ-phase matrix. It was deduced that these alloys had a potential use at 800C in energy structural applications.

A method was proposed[72] for constructing deformation-mechanism maps, for open-cell foams, by beginning with the deformation behavior of the constitutive material and taking account of the foam geometry. This revealed the predominant foam-deformation mechanisms as a function of temperature and applied-stress. The model was successfully applied to experimental results on open-cell pure-nickel foams.

Densification kinetics during the spark plasma-sintering of FeAl and NiAl powders were analyzed[73] by using creep parameters which were deduced from the densification data. Model predictions were presented in the form of deformation-mechanism maps which could predict the creep behavior on the basis of the densification data.

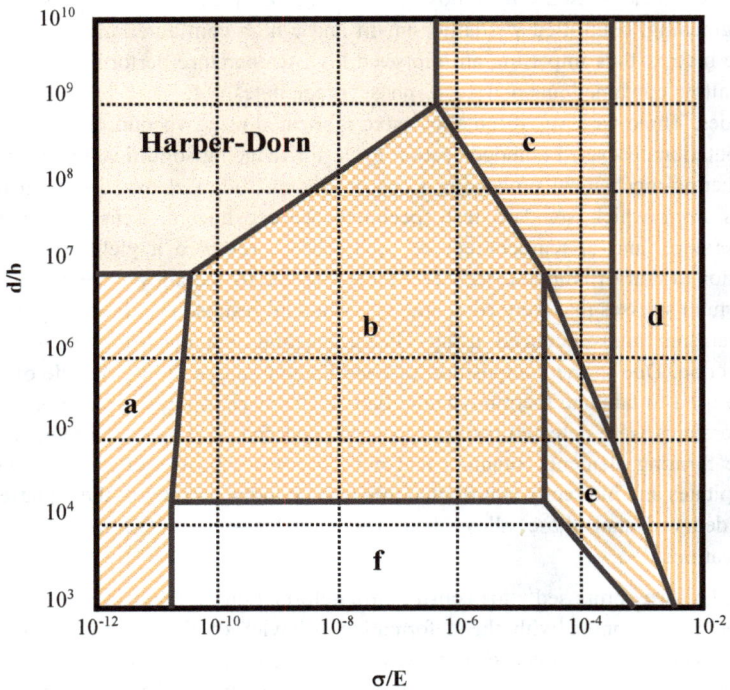

Figure 66. Deformation mechanism map for GH4742 nickel alloy; a – Coble diffusional flow controlled by grain-boundary diffusion and a stress exponent of 1, b – grain-boundary sliding controlled by lattice-diffusion and a stress exponent of 2, c – slip controlled by lattice-diffusion and a stress exponent of 5, d – slip controlled by lattice-diffusion and a stress exponent of 7, e – grain-boundary sliding controlled by pipe-diffusion and a stress exponent of 4, f – grain-boundary sliding controlled by grain-boundary diffusion and a stress exponent of 2

New experimental data were combined[74] with published data in order to construct high-temperature deformation-mechanism maps. These revealed a high-stress plasticity regime which was controlled by dislocation glide, a power-law creep field and a possible vacancy-diffusion creep field at low stresses. There was no sign of a grain-boundary diffusion creep field, but the curvature of temperature contours at high stress could be attributed to low-temperature dislocation-core diffusion-controlled power-law creep, or to power-law breakdown, or both. The vacancy-diffusion creep field was significant because extrapolation of the power-law creep data to low stresses could drastically underestimate the creep strains which might occur during the lifetime of an engineering component. The deformation maps were useful for summarizing data which spanned a wide range of experimental conditions. The deduction of parameters from the data in order to ensure that a change in mechanism occurred on predicted contours in the correct place put extra limits, on the analysis of results, which were not imposed if the two mechanisms were fitted separately.

Fifty-eight samples having average grain sizes ranging from 87.7 to 315μm were creep-tested[75] at 1023 to 1293K, using stresses of 14.1 to 105MPa, leading to the construction of a deformation-mechanism map. The data were well-represented by both high-temperature and low-temperature power-law creep mechanisms. The degree of influence of diffusion-based creep mechanisms, especially Coble creep, required further investigation. Extrapolation of the data in any direction out of the deformation-mechanism map was somewhat dubious.

The creep of GTD-111, with its high volume fraction of γ-phase, appears to involve several deformation mechanisms which operate under various combinations of temperature and stress. Their regions of operation can be described[76] in terms of a stress-temperature deformation-mechanism map. In order to construct the map, single-specimen creep tests were performed under constant-stress versus variable-temperature and constant-temperature versus variable-stress conditions. The map consisted of two separate dislocation-controlled mechanism fields: a stacking-fault plus antiphase boundary region and a dislocation-climb region. A diffusional creep mechanism operated at lower stresses.

On the basis of high-temperature compression tests, laws were determined[77] which related the strain-rate sensitivity index and the activation energy for deformation at various strain-rates and deformation temperatures. Dislocation evolution laws and deformation mechanisms, and the normalized flow stress were predicted by using deformation-mechanism maps.

Potassium

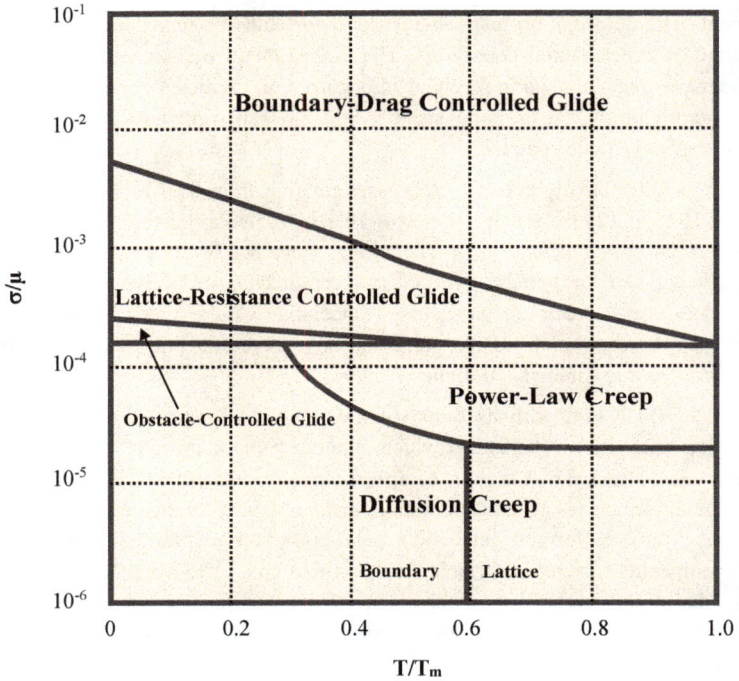

Figure 67. Deformation mechanism map for potassium with a grain size of 100 μm

The flow strength of potassium rises steeply, at homologous temperatures of less than 0.1, to a value of 5.4×10^{-3} at 0K[78]; a behavior which is typical of materials possessing a strong lattice resistance: that is, there exists an intrinsic resistance to dislocation motion which is caused by Peierls-Nabarro interaction of the dislocation with the lattice potential. A Bordoni internal friction peak which is due to double-kink formation has also been observed in potassium. The activation volume for the low-temperature plasticity of potassium was about $30b^3$ at a stress of about $\mu/1000$. This was indicative of lattice-resistance control.

Silver

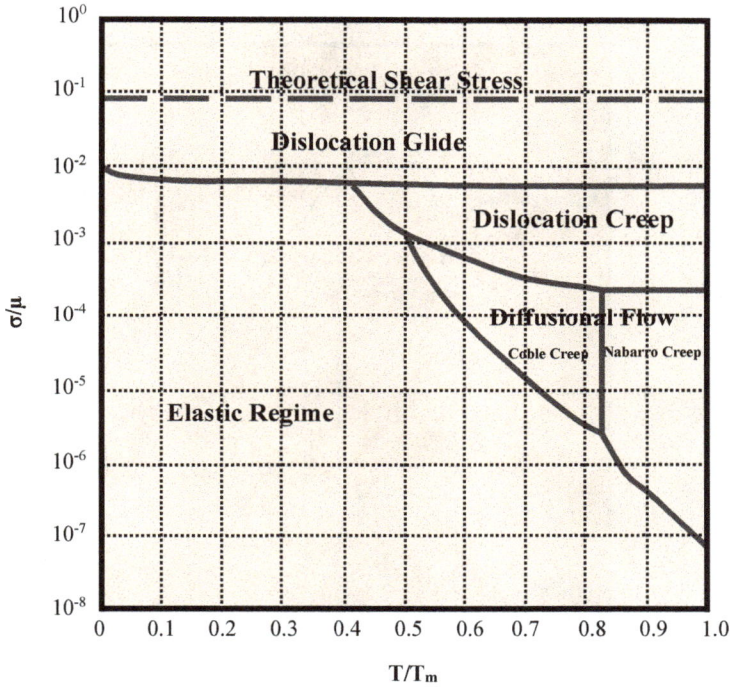

*Figure 68. Deformation mechanism map for silver,
with a grain size of 32 μm, at a critical strain rate of $10^{-8}/s$*

Silver typifies the behavior of the noble metals and of copper where, for all of these metals, the homologous temperature which separates Nabarro and Coble-creep is between 0.8 and 0.9. In the map for silver, an elastic field appears when the strain-rate is too small to be measurable. In constructing this map, the atomic volume was taken[79] to be $1.71 \times 10^{-23} cm^3$. The Burgers vector was $2.89 \times 10^{-8} cm$, the room-temperature shear modulus was $2.64 \times 10^{11} dyne/cm^2$ and the temperature-dependence of the shear modulus was $4.36 \times 10^{-4}/K$. The bulk and grain-boundary diffusivities were given by $D_l(cm^2/s) = 0.44\exp[-44.3(kcal/mol)/RT]$ and $D_{gb}(cm^2/s) = 0.09\exp[-21.5(kcal/mol)/RT]$, respectively.

Tin-

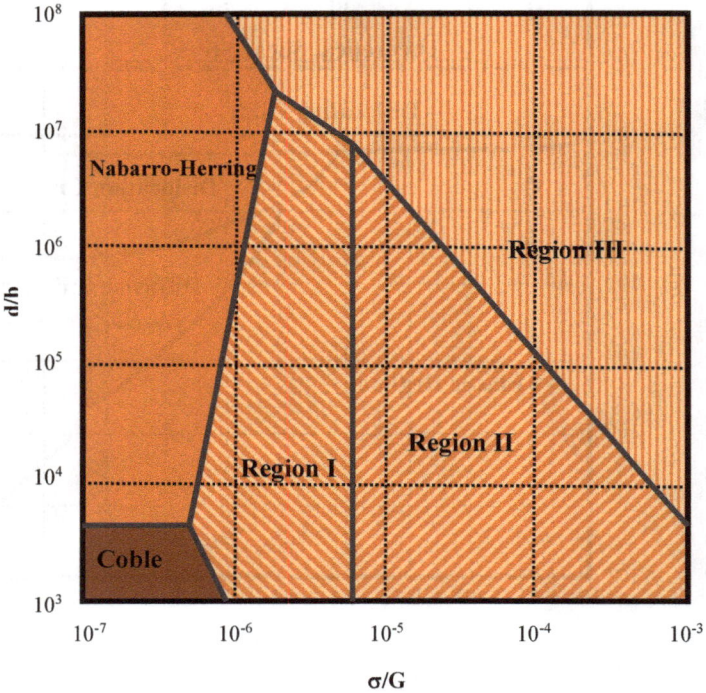

Figure 69. Deformation mechanism map for Sn–38%Pb alloy at 413K. Regions I, II and III denote regions of flow associated with sigmoidal, Nabarro–Herring and Coble diffusion creep

A Sn-38%Pb eutectic alloy containing 160ppm of antimony was processed[80] using equal-channel angular pressing at room temperature and tested in tension at 423K using strain-rates ranging from 10^{-4} to 10^{-1}/s. High superplastic elongations occurred at intermediate strain-rates, with a maximum elongation-to-failure of 2665%. The results were summarized by a deformation-mechanism map, with the normalized grain size being plotted against the normalized stress at 423K.

The alloy was also tested[81] in tension at a temperature of 413K. There was an increasing superplastic elongation with decreasing strain-rate, with elongations sometimes greater than 3000% at the lowest strain-rate. The experimental data were in excellent agreement with the predictions of the deformation-mechanism map.

The map provided[82] strong evidence that superplasticity occurred only at grain sizes which were sufficiently small, and a stable sub-grain structure was not formed during deformation.

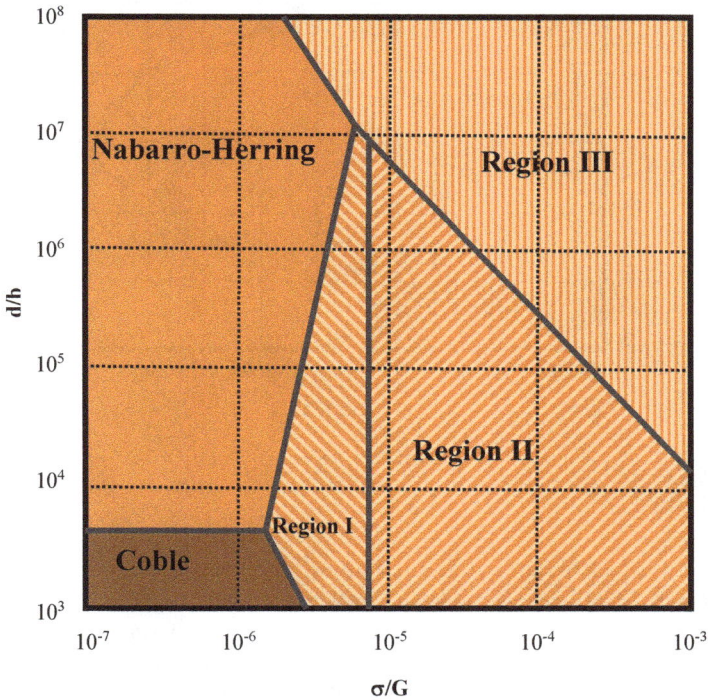

Figure 70. Deformation mechanism map for Sn-38%Pb at 423K. Regions I, II and III denote regions of flow associated with sigmoidal, Nabarro–Herring and Coble diffusion creep

Thallium

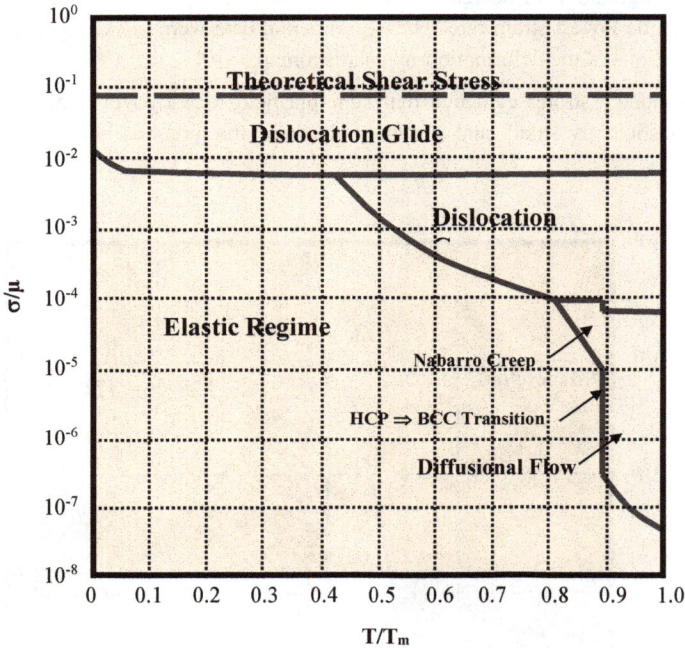

*Figure 71. Deformation mechanism map for thallium,
with a grain size of 32μm, at a critical strain rate of $10^{-8}/s$*

Hexagonal metals having non-ideal c/a ratios appear to differ from face-centered cubic metals in that they are more susceptible to Nabarro creep. This then enlarges the Nabarro-creep field at the expense of the Coble-creep field and pushes the transition temperature between them towards lower temperatures. Thallium is the only hexagonal metal with an almost ideal c/a ratio and does not exhibit the extended diffusional-creep field. In constructing this map, the atomic volume was taken[83] to be $2.92 \times 10^{-23} cm^3$. The Burgers vector was $3.36 \times 10^{-8} cm$, the room-temperature shear modulus was $0.55 \times 10^{11} dyne/cm^2$ and the temperature-dependence of the shear modulus was $17.0 \times 10^{-4}/K$. The bulk and grain-boundary diffusivities were given by $D_l(cm^2/s) = 0.7exp[-20(kcal/mol)/RT]$ and $D_{gb}(cm^2/s) = 1.0exp[-13.6(kcal/mol)/RT]$, respectively.

Titanium-

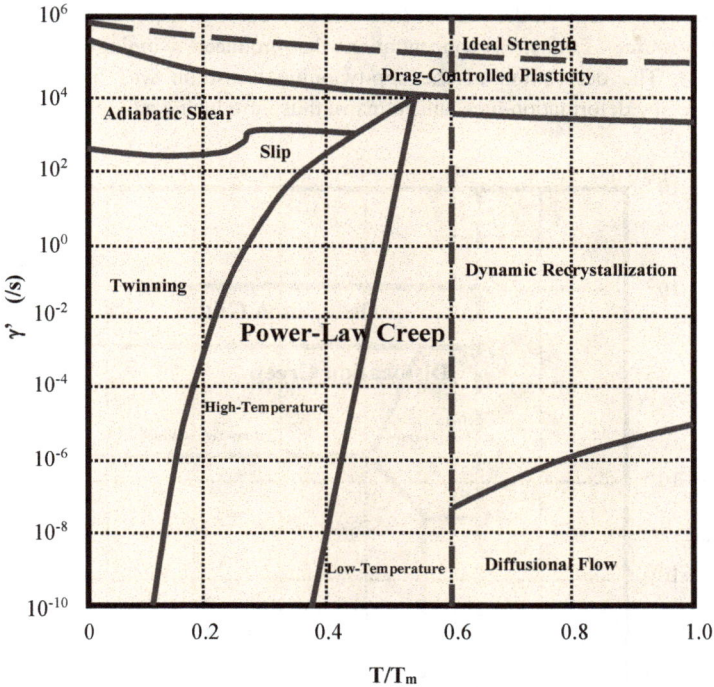

Figure 72. Deformation mechanism map for titanium with a grain size of 100μm. The vertical dashed line indicates the division between the low-temperature hexagonal close-packed α-phase and the body-centered cubic β-phase

The critical stress for twinning was predicted[84] as a function of temperature, strain-rate, grain size and stacking-fault energy. Plastic deformation via slip and twinning were considered to be competing mechanisms. The twinning stress was equated to a slip stress which was based upon plastic flow via the thermally-assisted movement of dislocations over obstacles, leading to a successful prediction of the slip-twinning transition. The model was applied to body-centered cubic, face-centered cubic and hexagonal close-packed metals and alloys. An expression for the twinning stress in body-centered cubic metals was developed on the basis of dislocation emission from a source and the formation of pile-ups as the rate-controlling mechanism. The critical size of the twin nucleus and the twinning stress were related to the twin-boundary energy and, in turn, to the stacking-fault energy in the case of face-centered cubic metals. Grain-scale pile-ups were not the source of the stress concentrations that produced twinning in face-centered cubic metals. The description of the slip-twinning transition was incorporated into Weertman-Ashby deformation-mechanism maps, thus introducing a twinning domain.

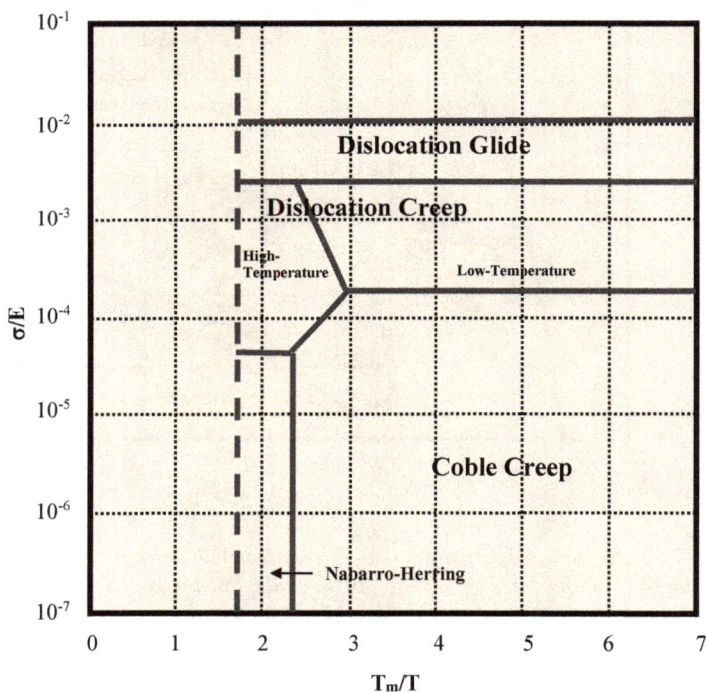

Figure 73. Redrawn original deformation mechanism map for titanium. The vertical dashed line marks the transition between the alpha (right) and beta (left) phases

Tensile deformation of single crystal titanium nanowires with sizes ranging from 3 to 20nm along [00•1] was investigated[85] using molecular dynamics simulations. The initial yielding at various strain-rates is induced by the nucleation of 10•2 twinning. Following saturation of the twin volume-fraction, a size-dependent transition of deformation mechanisms in twinned regions was observed. At strain-rates ranging from 10^8 to 10^9/s, following deformation-twinning, the phase transformation from hexagonal close-packed to face-centered cubic dominated the plastic deformation of titanium nanowires. By increasing the sample size to 20nm, the phase transformation could be replaced by prismatic dislocation slip. At strain rates ranging from 10^9 to 10^{10}/s, the critical size for the transition from phase transformation to full dislocation-slip decreased with applied strain-rate. With further increase in sample-size, following the saturation of 10•2 twins, the initial single-crystal nanowire transformed into nanocrystalline wire. The subsequent plastic deformation mechanism in nanocrystalline nanowire of large size changed from grain-boundary dominated deformation to the cooperation of grain-boundary deformation and dislocation activity.

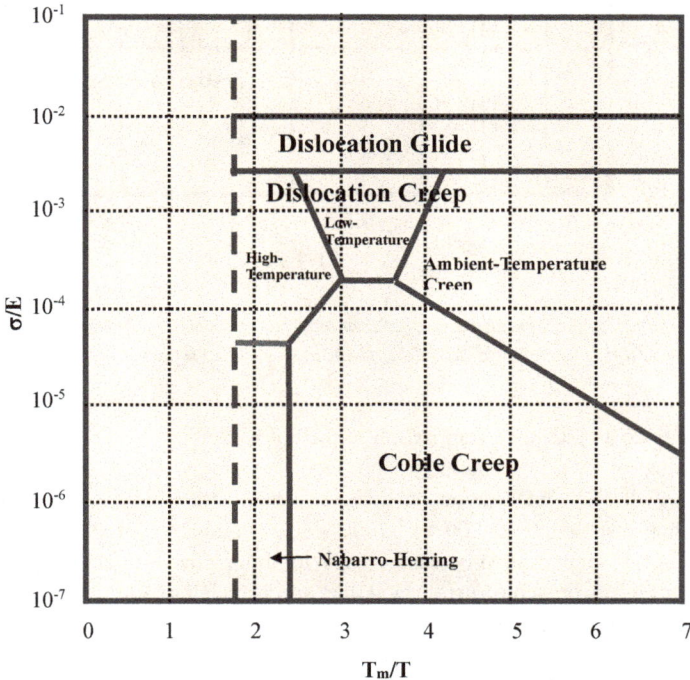

Figure 74. Redrawn original deformation mechanism map for titanium, including the ambient-temperature creep region. The vertical dashed line marks the transition between the alpha (right) and beta (left) phases

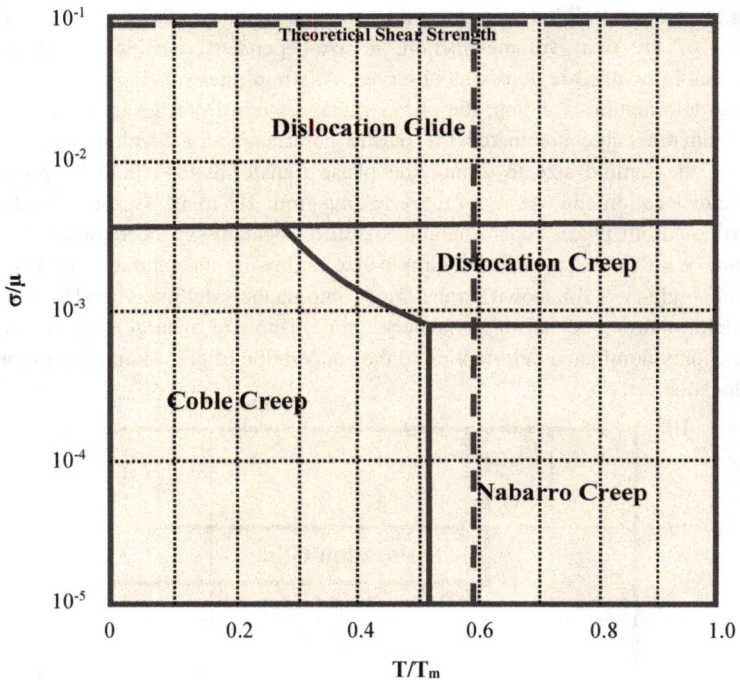

Figure 75. Deformation mechanism map for titanium with a grain size of 2μm. The vertical dashed line indicates the α-β phase transition

In order to make a systematic comparison of the hot-workability and to guide the further hot-processing of powder-metallurgy and ingot-metallurgy Ti-5Al-5V-5Mo-3Cr (Ti-5553) alloys, the hot deformation behavior and microstructural evolution of the two alloys were investigated[86] at 700 to 1100C using strain-rates of 0.001 to 10/s. The activation-energy maps and processing maps for both powder-metallurgy and ingot-metallurgy alloys were constructed, and the specific deformation mechanisms were identified for each processing region. The results showed that powder-metallurgy alloy had a lower deformation resistance, smaller activation energy and larger optimum processing windows than those of ingot-metallurgy alloy. The dynamic α-phase precipitation mechanisms in powder-metallurgy alloy were diffusional globularization and coarsening, rather than diffusionless shearing and fracturing in ingot-metallurgy alloy. The extensive dynamic recrystallization occurred at 900 to 1050C for powder-metallurgy alloy and at 1000 to 1100C for ingot-metallurgy alloy. The dynamic recrystallization process was dominated by discontinuous dynamic recrystallization for the powder-metallurgy alloy, but by continuous dynamic recrystallization for ingot-metallurgy alloy. The powder-metallurgy alloy had a smaller flow instability region than its ingot-metallurgy counterpart in the hot-processing map. Schematic deformation-mechanism maps were eventually developed for both powder-metallurgy and ingot-metallurgy Ti-5553.

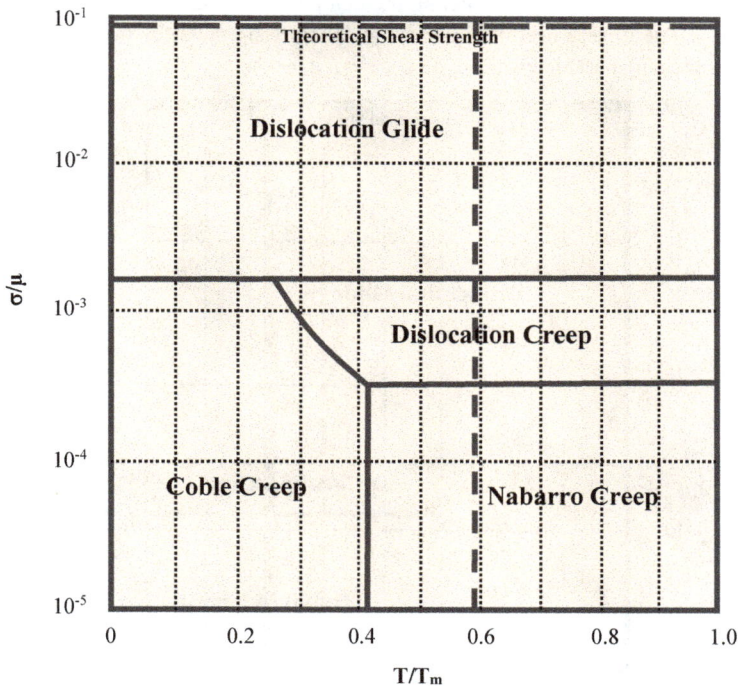

Figure 76. Deformation mechanism map for titanium with a grain size of 10μm. The vertical dashed line indicates the α-β phase transition

When metallic materials are subjected to stress, a number of independent or alternative mechanisms may initiate and contribute to deformation. In the case of γ-TiAl-based alloys, where the deformation kinetics are fairly complex, it is of particular interest to quantify the general constitutive relationship produced by each possible mechanism and to identify the predominant mechanism under any specific loading conditions.

For this purpose, Ashby-type deformation mechanism maps concerning six major deformation mechanisms were constructed[87] for various TiAl alloys having duplex and near-gamma microstructures. The general features as well as the effect of grain size on the appearance of the maps were analyzed. After a detailed discussion, it is believed that the proposed deformation mechanism maps are powerful tools in understanding the deformation mechanisms and predicting the deformation kinetics of duplex/near-gamma TiAl alloys. These were useful references for alloy design and for the determination of suitable processing parameters.

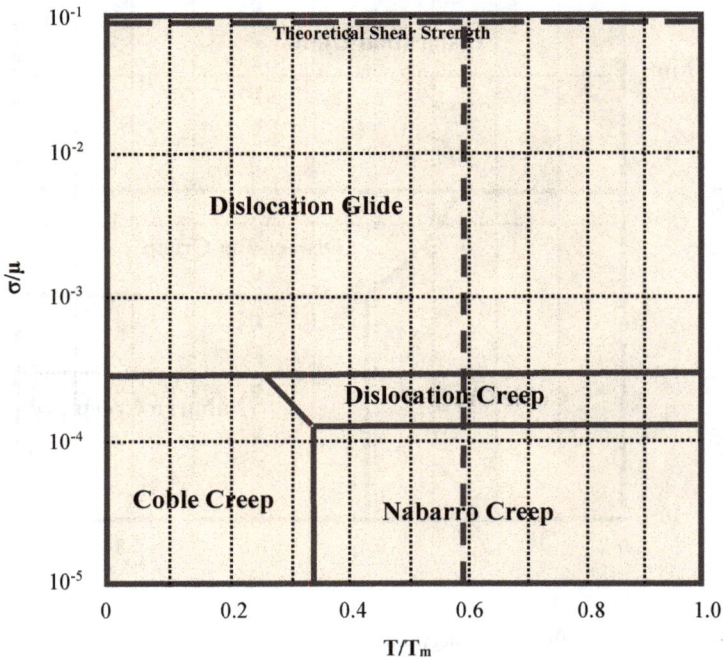

Figure 77. Deformation mechanism map for titanium with a grain size of 50μm. The vertical dashed line indicates the α-β phase transition

When seeking to obtain uniform and refined microstructures in Ti–6Al–4V alloy, the parameter loading path design plays a critical role. An attempt was made to separate dynamic recrystallization parameter domains from a chaotic parameter system. A succession of isothermal compression tests at 1023 to 1323K, at strain-rates ranging from 0.01 to 10/s, was performed in order to obtain basic computational data. Processing maps were constructed, and the mapping relationships between microstructural mechanisms and power dissipation efficiency were identified[88] by making microstructural observations.

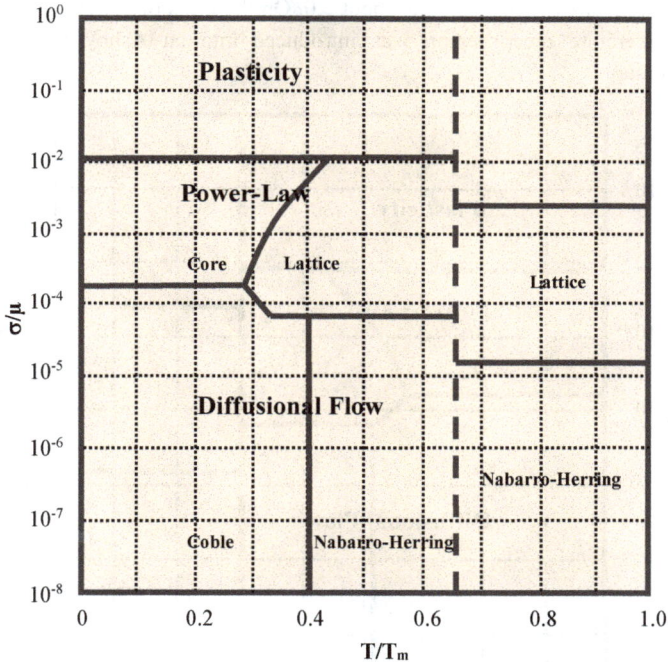

Figure 78. Deformation mechanism map for Ti-6wt%Al with a grain size of 100μm. The vertical dashed line separates the hexagonal close-packed α–phase (left) from the body-centered cubic β–phase (right)

The inner relationships between processing parameters and microstructural mechanisms were then mapped as a three-dimensional space in which the grain refinement parameter domains with a dynamic recrystallization mechanism were separated. These domains were however still chaotic, so finite-element simulations were used to scatter and clear the system. The optimum parameter loading paths were finally determined. Verification tests using different varying strain-rates were performed, and the grain refinement effect was more obvious than any constant strain-rate. One of the optimization results showed that the original grains, having an average size of 200μm, were refined to 9.7μm while, at a constant strain-rate, the original grains were refined to 22.3μm.

Creep behavior is exhibited[89] below the yield stress, by commercial-purity titanium, at ambient temperatures. The apparent activation energy was found to be about 10kJ/mol, and the apparent stress-exponent was about 5.0. On the basis of the creep parameters, an ambient-temperature creep region was introduced into an Ashby-type deformation-mechanism map.

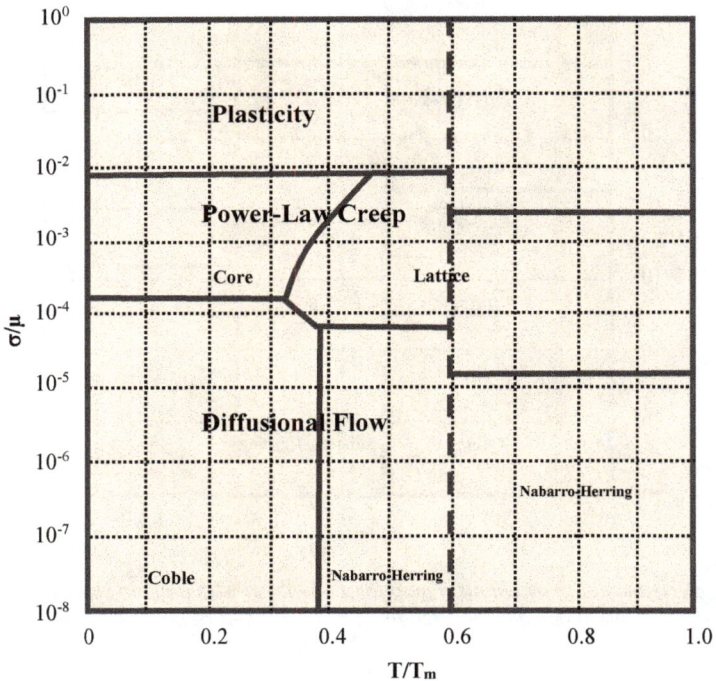

Figure 79. Deformation mechanism map for titanium with a grain size of 100μm. The vertical dashed line separates the hexagonal close-packed α–phase (left) from the body-centered cubic β–phase (right)

The constitutive equation in the ambient-temperature creep region of α-titanium was investigated[90] by performing creep tests on solute-strengthened and/or cold-rolled samples. Increasing the solute content and/or the thickness-reduction decreased the steady-state creep rate. The stress exponent increased with solute content, but was independent of thickness-reduction at low stresses. A micro-yielding stress, σ_{my}, was introduced in order to characterize the stress at which dislocations started to move. The stress exponent in the $\dot{\varepsilon}_s$ versus ($\sigma - \sigma_{my}$) graph became almost constant, with a value of 3, even for the solute-strengthened and/or cold-rolled titanium.

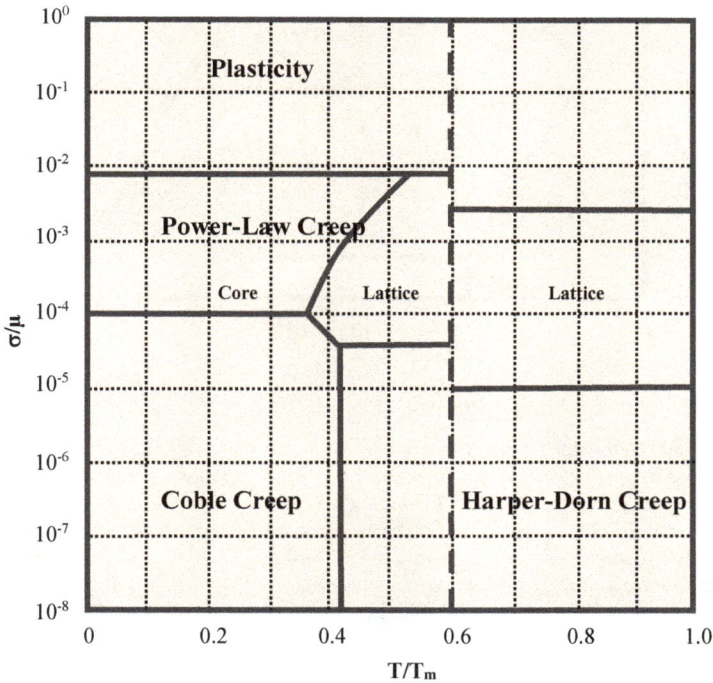

Figure 80. Deformation mechanism map for titanium with a grain size of 220μm. The vertical dashed line separates the hexagonal close-packed α–phase (left) from the body-centered cubic β–phase (right)

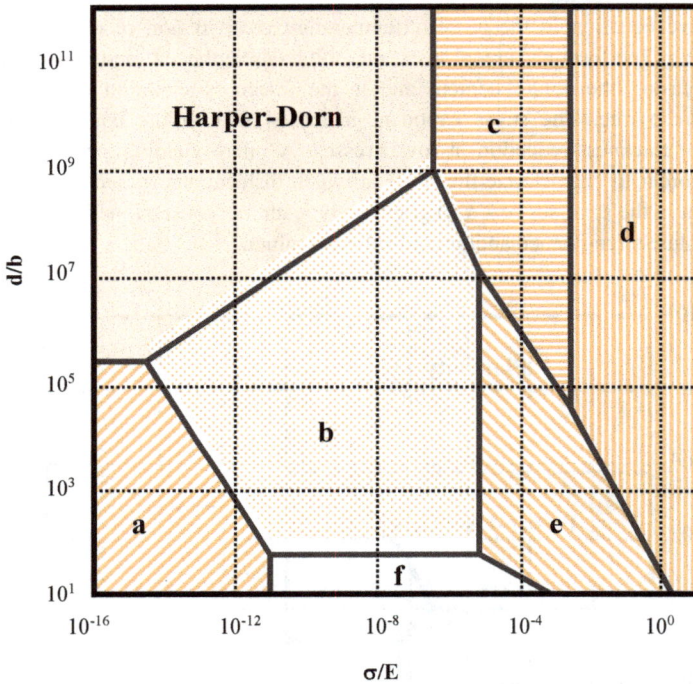

Figure 81. Deformation mechanism map for Ti-6Al-4V at 973K; a – diffusional flow controlled by grain-boundary diffusion and a stress exponent of 1, b – grain-boundary sliding controlled by lattice-diffusion and a stress exponent of 2, c – power-law slip controlled by lattice-diffusion and a stress exponent of 5, d – power-law slip controlled by lattice-diffusion and a stress exponent of 7, e – grain-boundary sliding controlled by grain-boundary diffusion and a stress exponent of 4, f – grain-boundary sliding controlled by grain-boundary diffusion and a stress exponent of 2

Hot-compression tests of as-cast TB6 titanium alloy were conducted at 948 to 1123K, using strain-rates of 0.001 to 10/s. The true stress-strain data were then used[91] to calculate the strain-rate sensitivity (m-value), power dissipation efficiency (η-value) and instability parameter (ξ-value) and the stable and unstable regimes were clearly delineated. The safe regions of the alloy were found to be: 948 to 1123K and 0.001 to 0.01/s, 1023 to 1123K and 0.01 to 0.1/s, 1036 to 1077K and 0.1 to 1/s. Deformation-mechanism maps were established on the basis of the parameters corresponding to various deformation mechanisms.

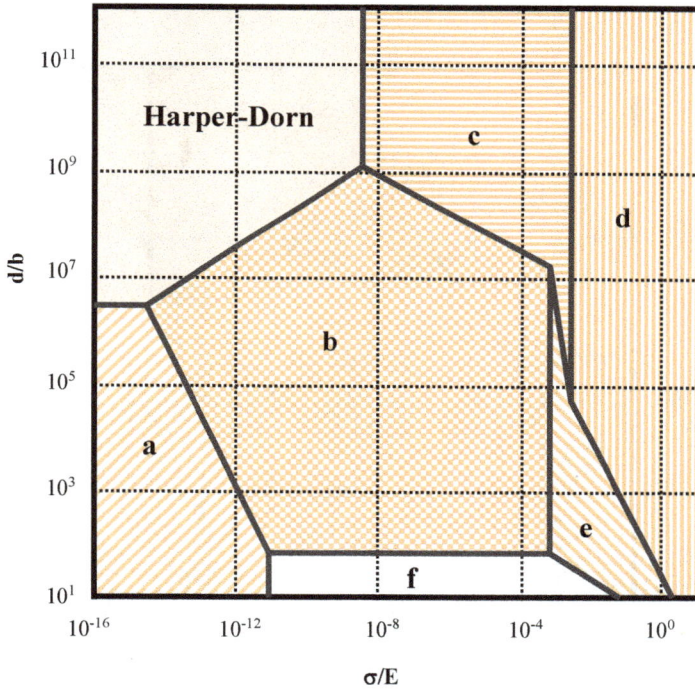

Figure 82. Deformation mechanism map for Ti-6Al-4V at 1123K; a – diffusional flow controlled by grain-boundary diffusion and a stress exponent of 1, b – grain-boundary sliding controlled by lattice-diffusion and a stress exponent of 2, c – power-law slip controlled by lattice-diffusion and a stress exponent of 5, d – power-law slip controlled by lattice-diffusion and a stress exponent of 7, e – grain-boundary sliding controlled by grain-boundary diffusion and a stress exponent of 4, f – grain-boundary sliding controlled by grain-boundary diffusion and a stress exponent of 2

The hot-workability of as-forged Ti-10V-2Fe-3Al alloy was evaluated[92]. The intrinsic relationships between deformation mechanisms and processing parameters were meanwhile determined by processing maps on the basis of a dynamic materials model, with the input stress-strain data being collected from a series of isothermal compression tests at temperatures of 948 to 1123 K (across the β-transus) and strain-rates of 0.001 to 10/s. For a set of discrete true strains, the response maps of strain-rate sensitivity exponent (m-value), power dissipation efficiency (η-value) and instability parameter (ξ-value) were again deduced.

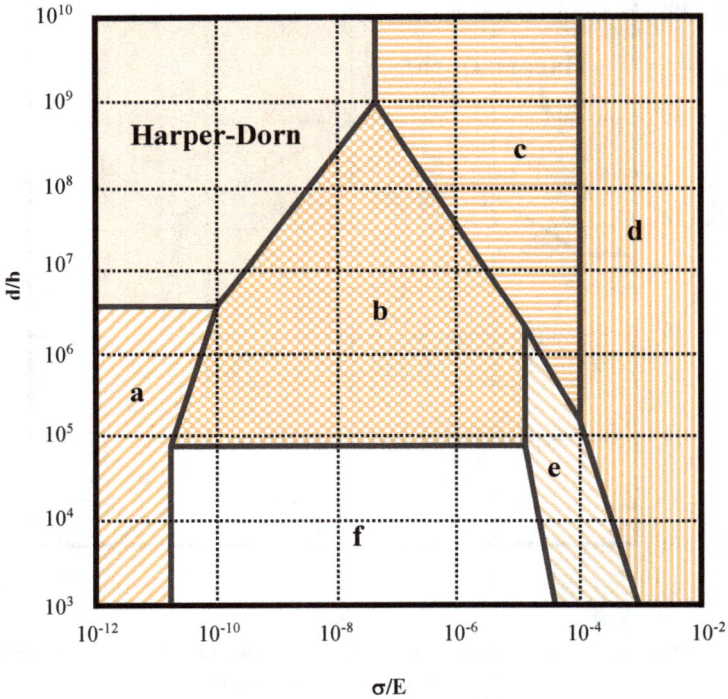

Figure 83. Deformation mechanism map for Ti-6Al-4V at 1103K; a – diffusional flow controlled by grain-boundary diffusion and a stress exponent of 1, b – grain-boundary sliding controlled by lattice-diffusion and a stress exponent of 2, c – slip controlled by lattice-diffusion and a stress exponent of 5, d – slip controlled by lattice-diffusion and a stress exponent of 7, e – grain-boundary sliding controlled by diffusion and a stress exponent of 4, f – grain-boundary sliding controlled by lattice-diffusion and a stress exponent of 2

A processing map which corresponded to each true strain was then constructed by superposing an instability map over a power dissipation map. According to the m-criterion, η-criterion and ξ-criterion, stable regions with higher power dissipation efficiency (η > 0.3) and unstable regimes with a negative strain-rate sensitivity exponent and instability parameter (m < 0 and ξ < 0) were identified. A deformation-mechanism map was established on the basis of which of the parameter domains corresponded to the possible deformation mechanisms.

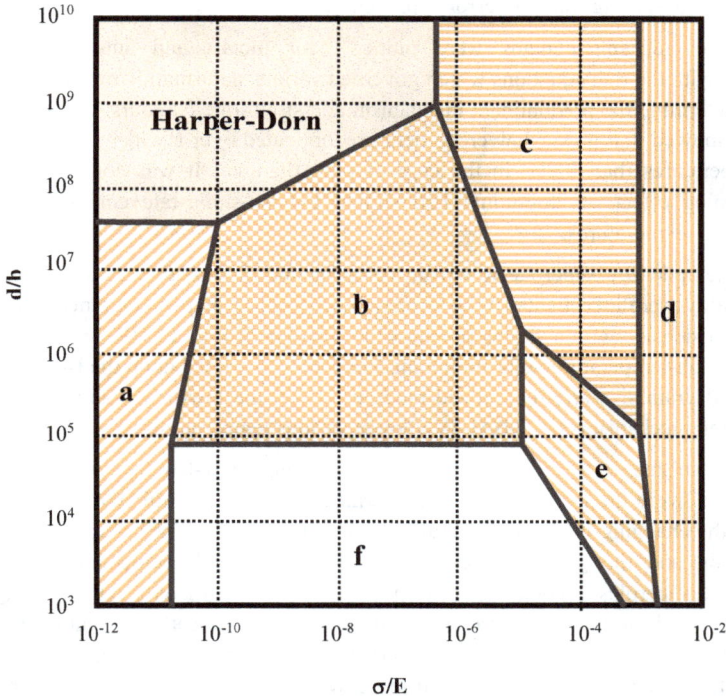

Figure 84. Deformation mechanism map for Ti-6Al-4V at 1163K; a – diffusional flow controlled by grain-boundary diffusion and a stress exponent of 1, b – grain-boundary sliding controlled by lattice-diffusion and a stress exponent of 2, c – slip controlled by lattice-diffusion and a stress exponent of 5, d – slip controlled by lattice-diffusion and a stress exponent of 7, e – grain-boundary sliding controlled by diffusion and a stress exponent of 4, f – grain-boundary sliding controlled by lattice-diffusion and a stress exponent of 2

Using the criteria of two main stable strain-softening mechanisms, i.e. globularization and dynamic recovery, the globularization-predominant parameter domain in the $\alpha+\beta$-phase temperature range and the dynamic recovery-predominant parameter domain in the β-phase temperature range were identified. Over a wide temperature range which crossed the β-transus, and for a wide strain-rate range, the clear separation of stable and unstable parameter regions which corresponded to different deformation mechanisms aided the design of various hot-forming processes for Ti-10V-2Fe-3Al alloy without needing the use of time-consuming trial-and-error approaches.

Deformation-mechanism maps were plotted[93] for metals and substitutional solid solutions, with the rate equations which governed various deformation mechanisms being used to determine the predominant mechanism and strain-rate contours. The map for Ti-6wt%Al showed that power-law creep here predominated over a wider range of stresses and temperatures than it did in the case of pure titanium. It was concluded that the strengthening effect of aluminium had shifted the iso-strain-rate contours to higher stresses and temperatures.

The low-temperature superplastic tensile behavior and deformation mechanisms of Ti-6Al-4V alloy were investigated[94]. Deformation-mechanism maps which incorporated the dislocation density within grains at temperatures of 973 and 1123K were drawn. By means of high-temperature deformation mechanism maps, based upon the Burgers-vector normalized grain size and normalized modulus stress, the low-temperature superplastic deformation mechanisms of Ti-6Al-4V alloy at various temperatures were deduced.

The mechanisms operating in Ti-6Al-4V during low-temperature superplastic deformation were studied[95] by using a thermocompression simulation machine. This clarified the changing forms of the strain-rate sensitivity index, the activation energy for deformation and the grain size at various strain-rates and temperatures. The low-temperature superplastic compression-forming zone and the rheological instability zone were analyzed in terms of hot-processing theories. The dislocation evolution laws and deformation mechanisms of grain size with Burgers vector normalization, and rheological stress with modulus normalization, during low-temperature superplastic compression were predicted by using deformation-mechanism maps.

Tungsten

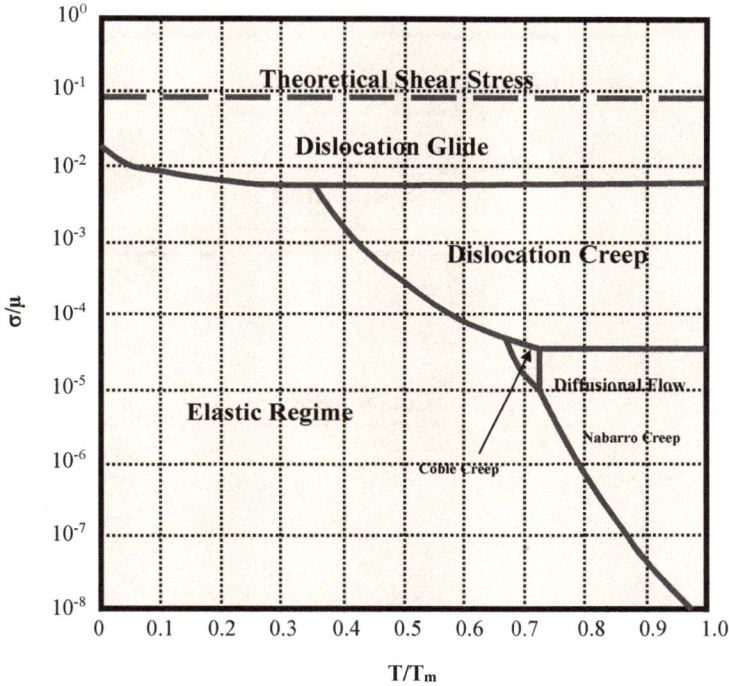

*Figure 85. Deformation mechanism map for tungsten
with a grain size of 32 μm, at a critical strain-rate of $10^{-8}/s$*

In constructing this map, the atomic volume was taken[96] to be 1.59 x 10^{-23}cm³. The Burgers vector was taken[97] to be 2.74 x 10^{-8}cm, the room-temperature shear modulus was 15.5 x 10^{11}dyne/cm² and the temperature-dependence of the shear modulus was 1.04 x 10^{-4}/K. The bulk and grain-boundary diffusivities were given by D_l(cm²/s) = 5.6exp[-140(kcal/mol)/RT] and D_{gb}(cm²/s) = 10exp[-90.5(kcal/mol)/RT], respectively.

Zinc-

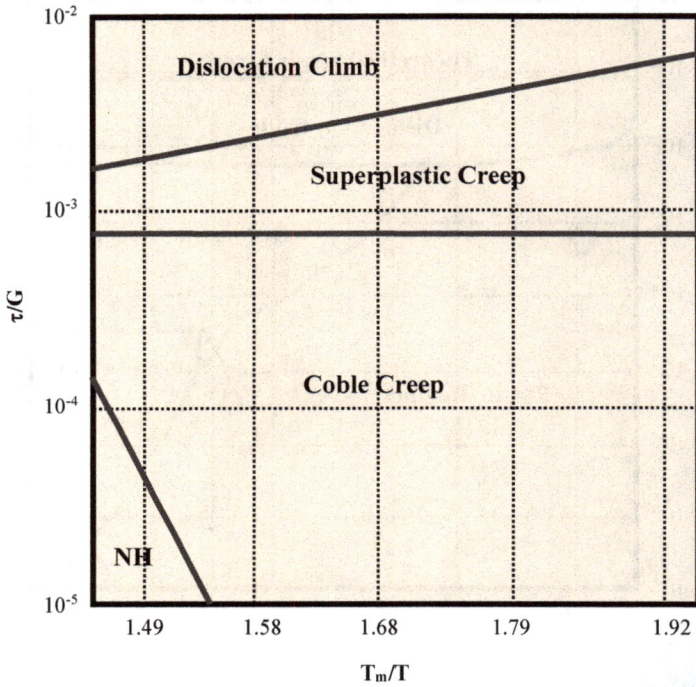

*Figure 86. Deformation mechanism map for Zn-22%Al
with a grain size of 1.7 x 10⁻⁴cm*

The number of dislocations and the dislocation density within a grain and at the grain boundary were incorporated[98] into the Mohamed-Kawasaki-Langdon deformation-mechanism map in an attempt to describe their distribution. It was found that, in the superplasticity (region II) regime, whether or not dislocations existed within the grain depended upon the Burgers vector normalized grain size and the shear modulus normalized stress. According to the maps, the number of dislocations at the grain boundary was about 20% of that within the grain. The dislocation forecasts for the Coble creep and dislocation-creep regimes were consistent with the predictions of the Coble and

Weertman mechanisms. A new map which included the dislocation density was proposed for Zn-22wt%Al eutectoid alloy, and its predictions were in good agreement with available experimental data.

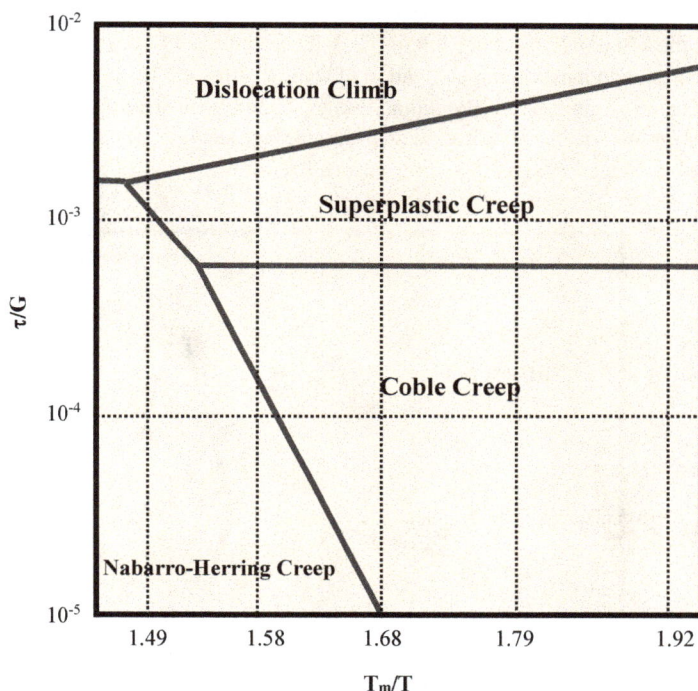

Figure 87. Deformation mechanism map for Zn-22%Al
with a grain size of 3.5 x 10-4cm

Severe plastic deformation is an attractive processing method for refining microstructures so as to have ultra-fine grain sizes within the sub-micron or even nanometre ranges. In severe plastic deformation processing, the most promising techniques are equal-channel angular pressing and high-pressure torsion. A conventional superplastic Zn-22%Al eutectoid alloy was processed[99] using both equal-channel angular pressing and high-pressure torsion. Experiments were aimed at demonstrating the evolution of hardness and microstructure and the enhancement of the superplastic properties of the Zn-Al alloy following processing using severe plastic deformation techniques. In addition, the flow mechanisms of the Zn-22% Al alloy were analyzed by using a deformation mechanism map.

A Zn-22% Al eutectoid alloy was processed[100] by equal-channel angular pressing in order to reduce the grain size to about 0.8μm. Tensile testing at 473K revealed superplastic characteristics, with a maximum elongation of some 2230% at a strain rate of 1.0 x 10^{-2}/s. The importance of grain-boundary sliding was evaluated by measuring the sliding offsets at adjacent grains via the displacement of surface marker lines on samples which had been pulled to elongations of 30% using various strain-rates.

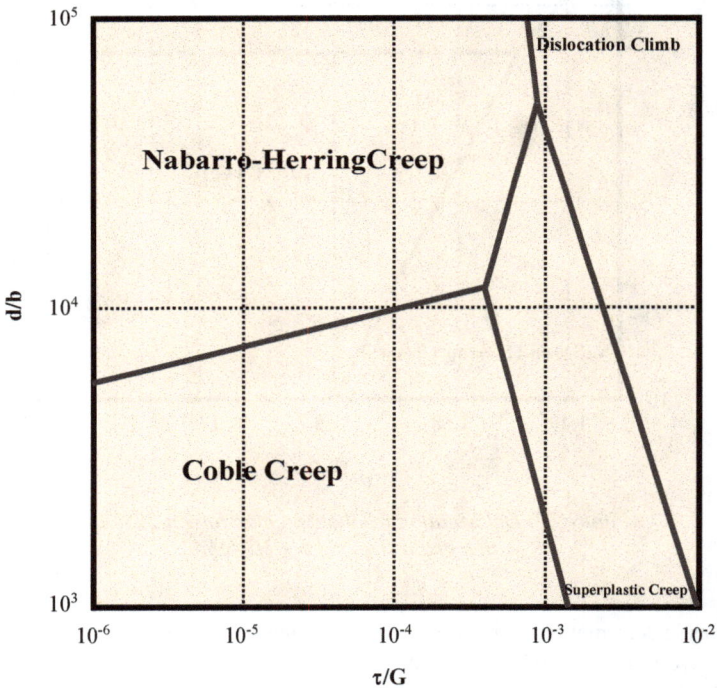

Figure 88. Deformation mechanism map for Zn-22%Al at 443K

The highest sliding contribution was recorded under test conditions which corresponded to the maximum superplastic ductility. There were relatively large offsets at the Zn|Zn and Zn|Al interfaces, but smaller offsets at the Al|Al interfaces. Analysis showed that the results were affected by the presence of agglomerates of similar grains which were present following equal-channel angular pressing processing, and especially by the increased fraction of Al|Al boundaries. The experimental results were in excellent agreement with the predictions of a deformation-mechanism map which depicted the flow behavior of the Zn-22%Al alloy, and the results confirmed the importance of grain-boundary sliding as being the predominant mechanism of superplastic flow following equal-channel angular pressing.

The relationships between the rate-controlling deformation mechanisms operating in Zn-22%Al eutectoid alloy were analyzed[101] using conventional deformation-mechanism maps. The latter were considered here to be lacking in generality in that many maps had to be constructed for many ranges of stress, temperature and grain size, in order to grasp the overall picture. In order to overcome this limitation, a 3-dimensional map was developed.

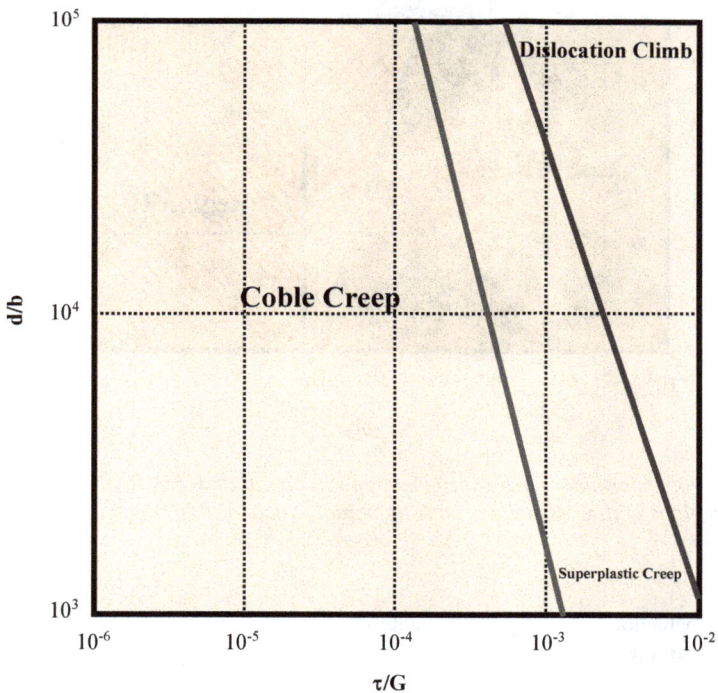

Figure 89. Deformation mechanism map for Zn-22%Al at 503K

Materials Research Forum LLC
https://doi.org/10.21741/9781644901694

The Zn-22%Al eutectoid alloy was processed[102] using high-pressure torsion at room temperature in order to produce an ultra-fine grain size of about 350nm. Tensile testing at 473K then revealed excellent superplastic properties, with elongations-to-failure of up to 1800% for a strain-rate of 10^{-1}/s. This was a record value of elongation, at the time, for high-pressure torsion treated material. The experimental data were in excellent agreement with a deformation-mechanism map constructed for a temperature of 473K.

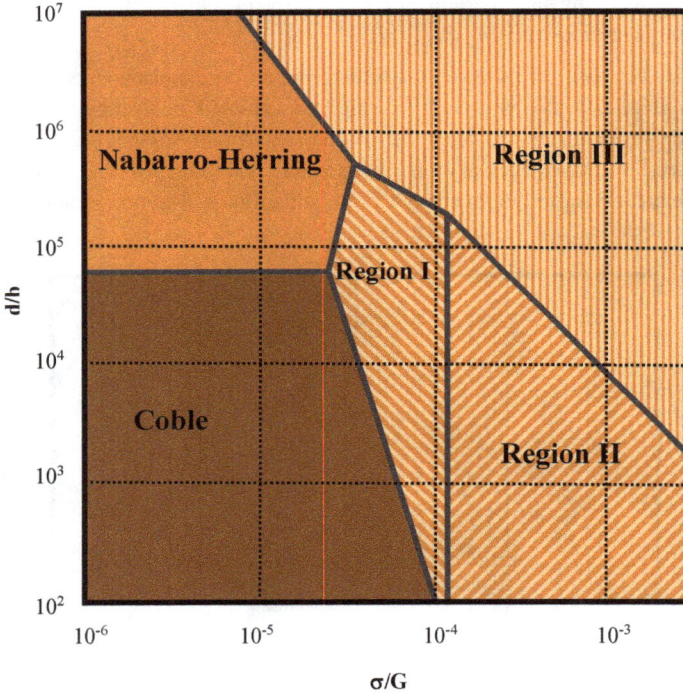

Figure 90. Deformation mechanism map of Zn–22%Al at 473K. Regions I, II and III denote regions of flow associated with sigmoidal, Nabarro–Herring and Coble diffusion creep

The creep behavior of polycrystalline materials can now be accurately described by using a set of equations which delineate the various mechanisms operating in intragranular and intergranular flow. The production of metals having sub-micron and nanometre grain sizes nevertheless questioned[103] the applicability of the flow processes to such materials. Their extension to samples having an ultra-fine grain size was explored.

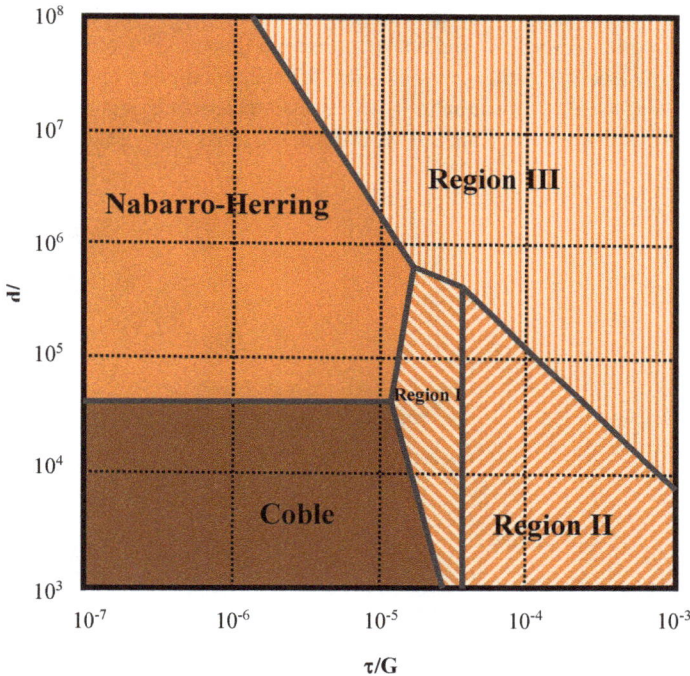

Figure 91. Deformation mechanism map for Zn-22%Al at 503K. Regions I, II and III denote regions of flow associated with sigmoidal, Nabarro–Herring and Coble diffusion creep

The characteristic features of superplastic deformation in the Zn-22%Al eutectoid alloy were summarized, and the results were presented[104] in the form of a deformation-mechanism map. A simple form of the map was presented which graphically illustrated the behavior of materials under steady-state creep conditions. Examples were given of maps for deformation mechanisms which occurred both independently and sequentially. The use of deformation-mechanism maps in the prediction of creep behavior was reviewed in detail, and an improved method was presented for estimating the total strain experienced following long-term exposure to stress and temperature. For a wide range of materials, the measured steady-state creep rates were essentially independent of the grain, size at large grain sizes, but significantly faster rates were often observed at small grain sizes. The various mechanisms which occur during high-temperature creep were reviewed, and it was shown that the increase at small grain sizes arises due to the occurrence of grain-boundary sliding, Nabarro-Herring and Coble creep. The predominance of various creep processes under different conditions of stress and grain size was demonstrated by constructing deformation-mechanism maps in which the normalized grain size was plotted against the normalized stress, at constant temperature.

Zirconium-

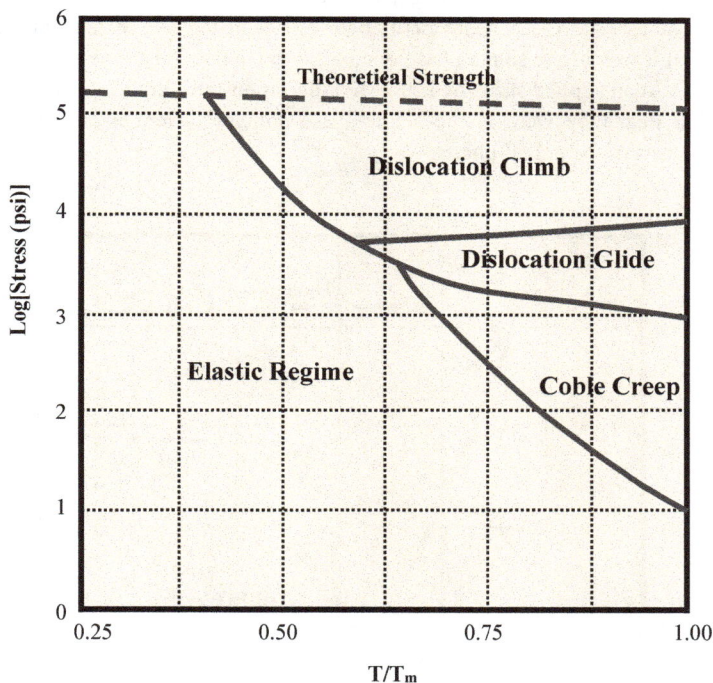

Figure 92. Deformation mechanism map for α-zirconium with a grain size of 60μm, at a strain-rate of $10^{-8}/s$. In this case, T_m is not a melting-point but the α-β transition temperature

The temperature of the α → β transformation for Zircaloy-4 and Zr-1%NbO alloys was determined[105] by using resistivity, calorimetric and thermodynamic calculations. The experimental and calculated results were in good agreement. The start temperature of the α ⇒ β transformation, calculated using the resistivity technique, was about 820C for Zircaloy-4 and 770C for Zr–1%NbO. A discrepancy between the measured and predicted start temperatures of the transformation was attributed to the so-called overheating effect upon the incubation time of diffusion-controlled phase transformations.

The steady-state creep behavior was determined for the single-phase and (α + β) temperature ranges, and the creep results were summarized by using deformation-mechanism maps. The creep behavior of both alloys in the α + β domain was more complex than it was in the single-phase domains. No simple constitutive equation could be identified. At an applied stress of 1 or 2MPa, the strain-rates in the α + β domain were much higher than they were in the single-phase domains; even in the case of high-temperature tests of the pure β-phase.

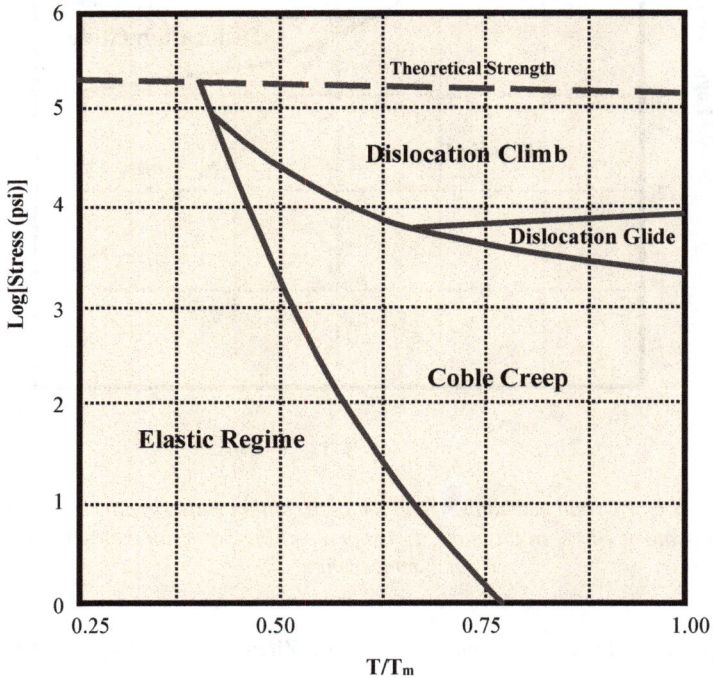

Figure 93. Deformation mechanism map for α-zirconium with a grain size of 10μm, at a strain-rate of 10^{-8}/s. In this case, T_m is not a melting-point but the α-β transition temperature

The observed creep exponent of near-unity suggested that, in the α + β domain, deformation was controlled by interphase interface sliding. Due to this complex behavior, only a power-law strain-rate dependence upon stress sufficed, with both the proportionality constant and the exponent being temperature-dependent. Upon comparing the alloys it was concluded that, in the absence of appreciable oxidation effects, the Zr–1%NbO alloy was more resistant to creep than was Zircaloy-4 in the high-temperature β-phase domain. It was a little less resistant in the α + β domain. In the single-phase domains, the creep behavior could be adequately described by a classic power-law equation.

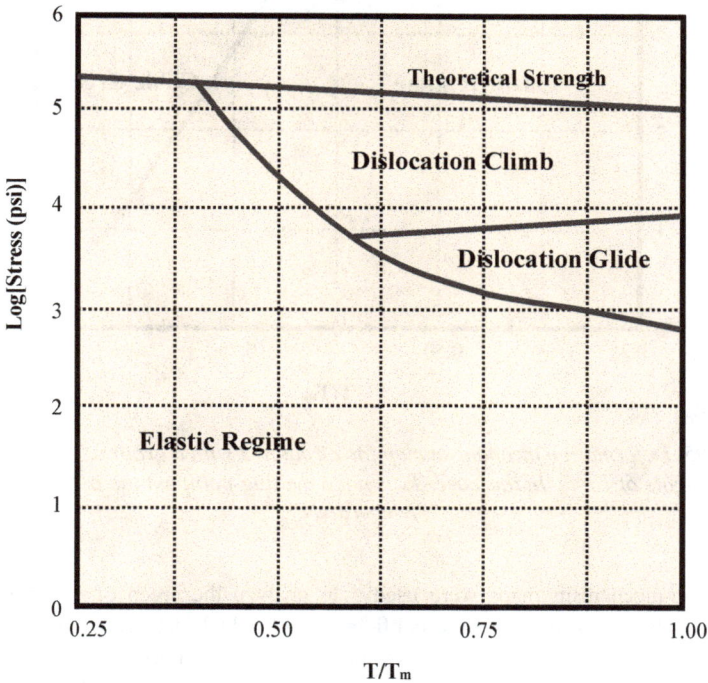

Figure 94. Deformation mechanism map for α-zirconium with a grain size of 300μm, at a strain-rate of 10^{-8}/s. In this case, T_m is not a melting-point but rather the α-β transition temperature

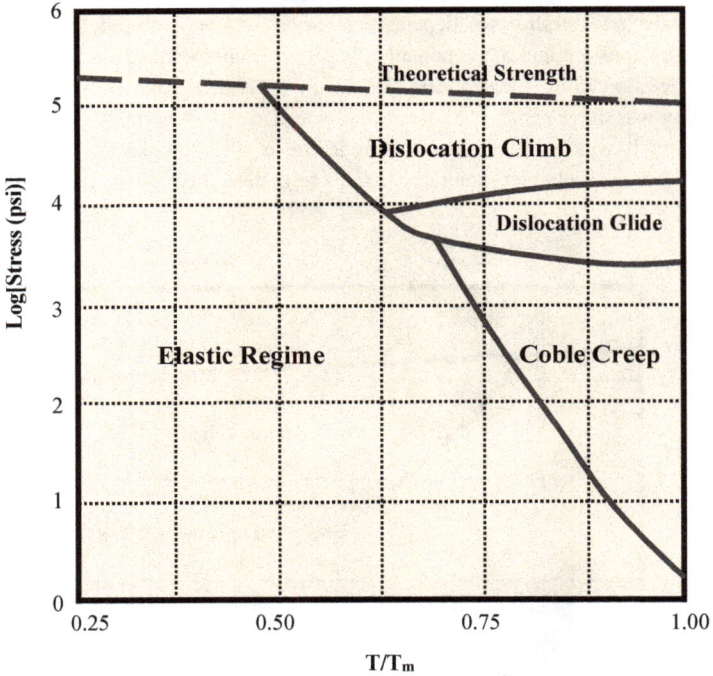

Figure 95. Deformation mechanism map for Zircaloy-2 with a grain size of 10μm, at a strain-rate of 10^{-8}/s. In this case, T_m is not a melting-point but the α-β transition temperature

Deformation-mechanism maps were used[106] to analyze the creep of α-zirconium and Zircaloy-2. The maps increased the confidence with which data could be extrapolated beyond the experimental range. The α-phase of zirconium exhibited unusual creep characteristics in that it deviated from the behavior observed in other metals and ceramics at high stresses. Deformation maps for α-zirconium were plotted for grain sizes of 10, 60 and 300μm. The experimental strain-rates were in good agreement with the rates predicted by the map. A predominant Coble creep was predicted to occur at low stresses in the deformation map for some grain sizes. No predominant Coble-creep regime was observed under similar experimental stress conditions at a grain-size of 300μm. This effect demonstrated the strong grain-size dependence of Coble creep. Data which were obtained below about 50% of the transition temperature did not agree well with the map. This was attributed to a possible temperature effect upon the diffusion coefficient. At stresses above about 10000psi there was a disagreement between the map and experiment. This was attributed to the possible occurrence of other high-stress dislocation glide mechanisms. A deformation map for Zircaloy-2 with a grain size of 10μm showed that it exhibited the same regions and general shape as the map for α-zirconium. An obvious difference between creep in α-zirconium and in Zircaloy-2 concerned the Coble diffusional flow regime.

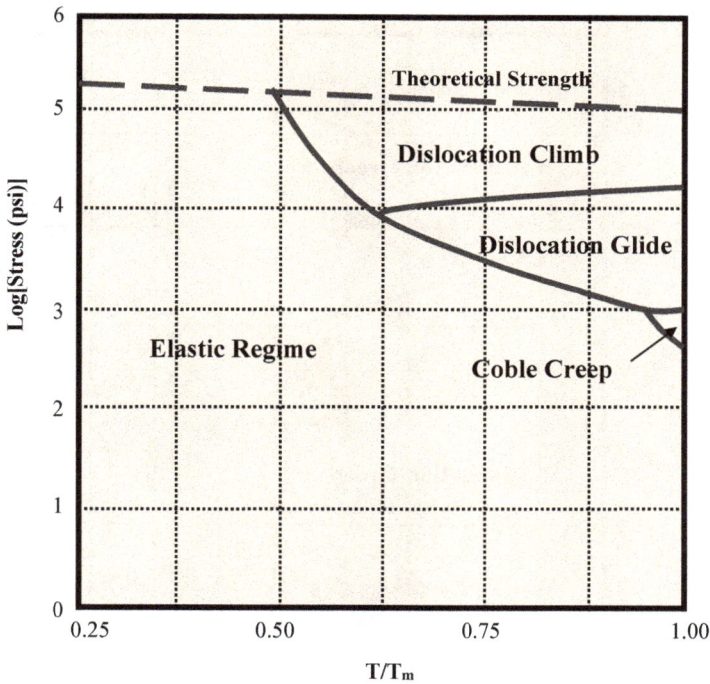

Figure 96. Deformation mechanism map for Zircaloy-2 with a grain size of 60μm, at a strain-rate of 10^{-8}/s. In this case, T_m is not a melting-point but the α-β transition temperature

The equivalent creep rates for Zircaloy-2 appeared to be shifted to higher stresses or to higher temperatures. At an homologous temperature of 0.825 a creep rate of 10^{-7}/s occurred at a stress level which was some two orders of magnitude higher in Zircaloy-2 than in zirconium. This was attributed to a difference in the grain-boundary diffusion coefficient, given that the grain-boundary diffusion activation energy was larger and the resultant diffusion coefficient was smaller in Zircaloy-2 at a given temperature. The higher activation energy appeared to be due to the addition of tin. The model which was applied to both zirconium and Zircaloy-2 appeared to break down at low temperatures and/or high stresses.

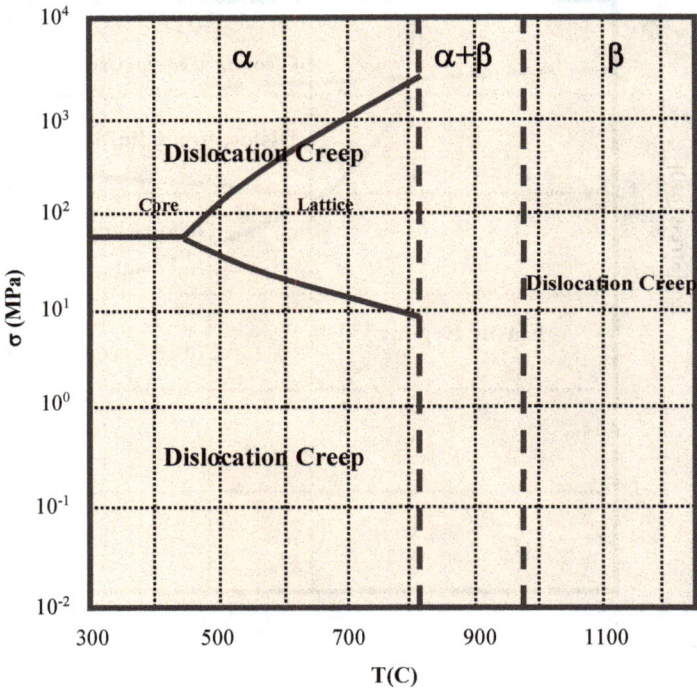

Figure 97. Deformation mechanism map for Zircaloy-4

Deformation maps for α-zirconium were plotted for grain sizes of 10, 60 and 300µm. The experimental strain-rates were in good agreement with the rates predicted by the map. A predominant Coble creep was predicted to occur at low stresses in the deformation map for some grain sizes. No predominant Coble-creep regime was observed under similar experimental stress conditions at a grain-size of 300µm. This effect demonstrated the strong grain-size dependence of Coble creep. Data which were obtained below about 50% of the transition temperature did not agree well with the map. This was attributed to a possible temperature effect upon the diffusion coefficient. At stresses above about 10000psi there was a disagreement between the map and experiment.

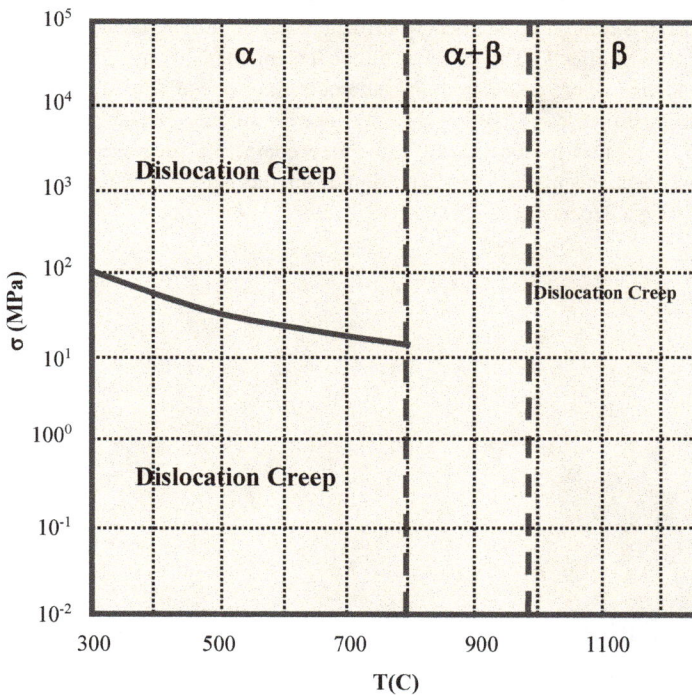

Figure 98. Deformation mechanism map for Zr–1%NbO

In typical coarse-grained alloys, the predominant plastic deformation mechanisms are dislocation gliding or climbing and material strengths can be tuned by modifying dislocation interactions with grain boundaries, precipitates, solid solutions and other defects. With a reduction in grain size, the increase in material strength obeys the classic Hall-Petch relationship up to the nano-grained material scale. Even at ambient temperatures, nano-grained materials exhibit strength-softening due to the so-called inverse Hall-Petch effect, as grain-boundary processes take over as being the predominant deformation mechanisms. On the other hand, at elevated temperatures, grain-boundary processes compete with grain-interior deformation mechanisms over a wide range of applied stresses and grain sizes. Rate-equation models and microstructure-based finite-element simulations have been compared. The latter explicitly accounts for grain-boundary sliding, grain-boundary diffusion and migration, as well as grain-interior dislocation creep effects. The explicit finite-element method has clear advantages for treating problems where microstructural heterogeneities play a critical role, such as in the gradient microstructures following shot-peening or welding. When combined with the Hall-Petch effect and its breakdown, the above competing processes help to construct deformation-mechanism maps by extension from the classic Frost-Ashby type to the those admitting a dependence on grain size.

Keyword Index

About the Author

Dr. Fisher has wide knowledge and experience of the fields of engineering, metallurgy and solid-state physics, beginning with work at Rolls-Royce Aero Engines on turbine-blade research, related to the Concord supersonic passenger-aircraft project, which led to a BSc degree (1971) from the University of Wales. This was followed by theoretical and experimental work on the directional solidification of eutectic alloys having the ultimate aim of developing composite turbine blades. This work led to a doctoral degree (1978) from the Swiss Federal Institute of Technology (Lausanne). He then acted for many years as an editor of various academic journals, in particular *Defect and Diffusion Forum*. In recent years he has specialized in writing monographs which introduce readers to the most rapidly developing ideas in the fields of engineering, metallurgy and solid-state physics. He is co-author of the widely-cited student textbook, *Fundamentals of Solidification*. Google Scholar credits him with 8187 citations and a lifetime h-index of 17.

References

[1] Prasad, Y.V.R.K., Rao, K.P., Sasidhara, S., *Hot Working Guide - a Compendium of Processing Maps*, ASM International, 2015.

[2] Weertman, J., Weertman, J.R. *Physical Metallurgy*, North-Holland 1965, 793.

[3] Weertman, J., Transactions of the ASM, 61, 1968, 681.

[4] Wang, J.N., Nieh, T.G., Scripta Metallurgica et Materialia, 33[4] 1995, 633-638.

[5] Wang, J.N., Wu, J.S., Ding, D.Y., Materials Science and Engineering A, 334[1-2] 2002, 275-279. https://doi.org/10.1016/S0921-5093(01)01896-2

[6] Ashby, M.F., Acta Metallurgica, 20[7] 1972, 887-897. https://doi.org/10.1016/0001-6160(72)90082-X

[7] Langdon, T.G., Metals Forum, 1, 1978, 59.

[8] Koleshko, V.M., Belitsky, V.F., Kiryushin, I.V., Thin Solid Films, 142[2] 1986, 199-212. https://doi.org/10.1016/0040-6090(86)90005-2

[9] Lüthy, H., White, R.A., Sherby, O.D., Materials Science and Engineering, 39[2] 1979, 211-216. https://doi.org/10.1016/0025-5416(79)90060-0

[10] Matsunaga, T., Ueda, S., Sato, E., Scripta Materialia, 63[5] 2010. 516-519. https://doi.org/10.1016/j.scriptamat.2010.05.019

[11] Kawasaki, M., Langdon, T.G., Journal of Materials Research, 28[13] 2013, 1827-1834. https://doi.org/10.1557/jmr.2013.55

[12] Langdon, T.G., Mohamed, F.A., Materials Science and Engineering, 32[2] 1978, 103-112. https://doi.org/10.1016/0025-5416(78)90029-0

[13] Samuelsson, L.C.A., Melton, K.N., Edington, J.W., Acta Metallurgica, 24[11] 1976, 1017-1026. https://doi.org/10.1016/0001-6160(76)90132-2

[14] Mohamed, F.A., Langdon, T.G., Metallurgical Transactions, 5[11] 1974, 2339-2345. https://doi.org/10.1007/BF02644014

[15] Kawasaki, M., Langdon, T.G., Acta Physica Polonica A, 128[4] 2015, 470-478. https://doi.org/10.12693/APhysPolA.128.470

[16] Mohamed, F.A., Langdon, T.G., Scripta Metallurgica, 9[2] 1975, 137-140. https://doi.org/10.1016/0036-9748(75)90582-7

[17] Koike, J., Utsunomiya, S., Shimoyama, Y., Maruyama, K., Oikawa, H., Journal of Materials Research, 13[11] 1998, 3256-3264. https://doi.org/10.1557/JMR.1998.0442

[18] Cao, F., Li, Z., Zhang, N., Ding, H., Yu, F., Zuo, L., Materials Science and Engineering A, 571, 2013, 167-183. https://doi.org/10.1016/j.msea.2013.02.010

[19] Xie, D.G., Zhang, R.R., Nie, Z.Y., Li, J., Ma, E., Li, J., Shan, Z.W., Acta Materialia, 188, 2020, 570-578. https://doi.org/10.1016/j.actamat.2020.02.013

[20] Xue, Y., Chen, S., Zhang, Z., Wang, Q., Yan, J., Journal of Materials Science, 54[10] 2019, 7908-7921. https://doi.org/10.1007/s10853-019-03355-5

[21] Ashby, M.F., Acta Metallurgica, 20[7] 1972, 887-897. https://doi.org/10.1016/0001-6160(72)90082-X

[22] Gao, Y.F., Yang, B., Nieh, T.G., Acta Materialia, 55[7] 2007, 2319-2327. https://doi.org/10.1016/j.actamat.2006.11.027

[23] Chokshi, A.H., Materials Chemistry and Physics, 210, 2018, 152-161. https://doi.org/10.1016/j.matchemphys.2017.07.079

[24] Brown, S.G.R., Evans, R.W., Wilshire, B., Materials Science and Technology, 3[1] 1987, 23-27. https://doi.org/10.1179/mst.1987.3.1.23

[25] Liu, Z., Wang, H., Haché, M.J.R., Chu, X., Irissou, E., Zou, Y., Acta Materialia, 193, 2020, 191-201. https://doi.org/10.1016/j.actamat.2020.04.041

[26] Matsunaga, T., Ueda, S., Sato, E., Scripta Materialia, 63[5] 2010. 516-519. https://doi.org/10.1016/j.scriptamat.2010.05.019

[27] Lee, K.L., Journal of Materials Science, 39[9] 2004, 3047-3055. https://doi.org/10.1023/B:JMSC.0000025831.58057.52

[28] Thouless, M.D., Gupta, J., Harper, J.M.E., Journal of Materials Research, 8[8] 1993, 1845-1852. https://doi.org/10.1557/JMR.1993.1845

[29] Brückner, W., Physica Status Solidi A, 176[2] 1999, 919-924. https://doi.org/10.1002/(SICI)1521-396X(199912)176:2<919::AID-PSSA919>3.0.CO;2-T

[30] Sekiguchi, A., Koike, J., Maruyama, K., Journal of the Japan Institute of Metals, 64[5] 2000, 379-382. https://doi.org/10.2320/jinstmet1952.64.5_379

[31] Ashby, M.F., Acta Metallurgica, 20[7] 1972, 887-897. https://doi.org/10.1016/0001-6160(72)90082-X

[32] Nishiyama, N., Wang, Y., Rivers, M.L., Sutton, S.R., Cookson, D., Geophysical Research Letters, 34[23] 2007, L23304. https://doi.org/10.1029/2007GL031431

[33] Chawake, N., Koundinya, N.T.B.N., Srivastav, A.K., Kottada, R.S., Scripta Materialia, 107, 2015, 63-66. https://doi.org/10.1016/j.scriptamat.2015.05.021

[34] Kloc, L., Sklenička, V., Materials Science and Engineering A, 234-236, 1997, 962-965. https://doi.org/10.1016/S0921-5093(97)00364-X

[35] Taleff, E.M., Kim, W.J, Sherby, O.D., TMS Annual Meeting, 1998, 209-218.

[36] Ogata, T., Materials at High Temperatures, 27[1] 2010, 11-19.
https://doi.org/10.3184/096034009X12602021834640

[37] Whittaker, M.T., Wilshire, B., Metallurgical and Materials Transactions A, 44[S1]
2013, S136-S153. https://doi.org/10.1007/s11661-012-1160-2

[38] Monteiro, S.N., Margem, F.M., Bolzan, L.T., Fernandes, G.L.N., Candido, V.S.,
Materials Science Forum, 869, 2016, 543-549.
https://doi.org/10.4028/www.scientific.net/MSF.869.543

[39] Maruyama, K., Sawada, K., Koike, J., Sato, H., Yagi, K., Materials Science and
Engineering A, 224[1-2] 1997, 166-172. https://doi.org/10.1016/S0921-5093(96)10566-9

[40] Priest, R.H., Ellison, E.G., Materials Science and Engineering, 49[1] 1981, 7-17.
https://doi.org/10.1016/0025-5416(81)90128-2

[41] Bano, N., Koul, A.K., Nganbe, M., Metallurgical and Materials Transactions A,
45[4] 2014, 1928-1936. https://doi.org/10.1007/s11661-013-2172-2

[42] Takahashi, Y., Yamane, T., Journal of Materials Science, 16[11] 1981, 3171-3182.
https://doi.org/10.1007/BF00540326

[43] Zauter, R., Petry, F., Christ, H.J., Mughrabi, H., Materials Science and Engineering
A, 124[2] 1990, 125-132. https://doi.org/10.1016/0921-5093(90)90142-P

[44] Langdon, T.G., Mohamed, F.A., Journal of Materials Science, 13[6] 1978, 1282-
1290. https://doi.org/10.1007/BF00544735

[45] Miller, D.A., Langdon, T.G., Metallurgical Transactions A, 10, 1979, 1635-1641.
https://doi.org/10.1007/BF02811696

[46] Murakami, M., Thin Solid Films, 55[1] 1978, 101-111.
https://doi.org/10.1016/0040-6090(78)90078-0

[47] Ashby, M.F., Acta Metallurgica, 20[7] 1972, 887-897. https://doi.org/10.1016/0001-
6160(72)90082-X

[48] Chen, Z., Cai, H., Zhang, X., Wang, F., Tan, C., Science in China E, 49[5] 2006,
521-536. https://doi.org/10.1007/s11431-006-2016-z

[49] Matsunaga, T., Ueda, S., Sato, E., Scripta Materialia, 63[5] 2010, 516-519.
https://doi.org/10.1016/j.scriptamat.2010.05.019

[50] Kim, W.J., Chung, S.W., Chung, C.S., Kum, D., Acta Materialia, 49[16] 2001,
3337-3345. https://doi.org/10.1016/S1359-6454(01)00008-8

[51] Kim, W.J., Park, J.D., Yoon, U.S., Journal of Alloys and Compounds, 464[1-2]
2008, 197-204. https://doi.org/10.1016/j.jallcom.2007.10.067

[52] Bisht, A., Kamble, N., Kumar, L., Avadhani, G.S., Roy, A., Silberschmidt, V., Suwas, S., International Journal of Materials Research, 110[6] 2019, 524-533. https://doi.org/10.3139/146.111766

[53] Sim, G.D., Kim, G., Lavenstein, S., Hamza, M.H., Fan, H., El-Awady, J.A., Acta Materialia, 144, 2018, 11-20. https://doi.org/10.1016/j.actamat.2017.10.033

[54] Cao, F.R., Guan, R.G., Ding, H., Li, Y.L., Zhou, G., Cui, J.Z., Advanced Materials Research, 189-193, 2011, 2504-2510. https://doi.org/10.4028/www.scientific.net/AMR.189-193.2504

[55] Hou, Q.Y., Wang, J.T., Advanced Materials Research, 146-147, 2011, 225-232. https://doi.org/10.4028/www.scientific.net/AMR.146-147.225

[56] Paufler, P., Kristall und Technik, 13[5] 1978, 587-590. https://doi.org/10.1002/crat.19780130522

[57] Chung, S.W., Higashi, K., Kim, W.J., Materials Science and Engineering A, 372[1-2] 2004, 15-20. https://doi.org/10.1016/j.msea.2003.08.125

[58] Kim, W.J., Scripta Materialia, 58[8] 2008, 659-662. https://doi.org/10.1016/j.scriptamat.2007.11.038

[59] Cao, F.R., Ding, H., Li, Y.L., Zhou, G., Chinese Journal of Nonferrous Metals, 19[11] 2009, 1908-1916.

[60] Figueiredo, R.B., Langdon, T.G., Materials Science and Engineering A, 787, 2020, 139489. https://doi.org/10.1016/j.msea.2020.139489

[61] Kawasaki, M., Figueiredo, R.B., Langdon, T.G., Materials Science Forum, 879, 2017, 48-53. https://doi.org/10.4028/www.scientific.net/MSF.879.48

[62] Cao, F.R., Ding, H., Li, Y.L., Zhou, G., Cui, J.Z., Materials Science and Engineering A, 527[9] 2010, 2335-2341. https://doi.org/10.1016/j.msea.2009.12.029

[63] Li, H., Liang, Y., Zhao, L., Hu, J., Han, S., Lian, J., Journal of Alloys and Compounds, 709, 2017, 566-574. https://doi.org/10.1016/j.jallcom.2017.03.188

[64] Detrois, M., McCarley, J., Antonov, S., Helmink, R.C., Goetz, R.L., Tin, S., Materials at High Temperatures, 33[4-5] 2016, 310-317. https://doi.org/10.1080/09603409.2016.1155689

[65] Smith, T.M., Unocic, R.R., Deutchman, H., Mills, M.J., Materials at High Temperatures, 33[4-5] 2016, 372-383. https://doi.org/10.1080/09603409.2016.1180858

[66] McCarley, J., Helmink, R., Goetz, R., Tin, S., Metallurgical and Materials Transactions A, 48[4] 2017, 1666-1677. https://doi.org/10.1007/s11661-017-3977-1

[67] Mello, A.W., Nicolas, A., Sangid, M.D., Materials Science and Engineering A, 695, 2017, 332-341. https://doi.org/10.1016/j.msea.2017.04.002

[68] Detrois, M., Rotella, J., Hardy, M., Tin, S., Sangid, M.D., Integrating Materials and Manufacturing Innovation, 6[4] 2017, 265-278. https://doi.org/10.1007/s40192-017-0103-6

[69] Barat, K., Ghosh, M., Sivaprasad, S., Kar, S.K., Tarafder, S., Metallurgical and Materials Transactions A, 49[10] 2018, 5211-5226. https://doi.org/10.1007/s11661-018-4760-7

[70] Song, X.T., Guo, R.P., Wang, Z., Wang, X.J., Yang, H.J., Qiao, J.W., Han, L.N., Liaw, P.K., Wu, Y.C., Intermetallics, 114, 2019, 106591. https://doi.org/10.1016/j.intermet.2019.106591

[71] Detrois, M., Jablonski, P.D., Antonov, S., Li, S., Ren, Y., Tin, S., Hawk, J.A., Journal of Alloys and Compounds, 792, 2019, 550-560. https://doi.org/10.1016/j.jallcom.2019.04.054

[72] Zurob, H.S., Bréchet, Y., Journal of Materials Science, 40[22] 2005, 5893-5901. https://doi.org/10.1007/s10853-005-5031-8

[73] Chawake, N., Koundinya, N.T.B.N., Srivastav, A.K., Kottada, R.S., Scripta Materialia, 107, 2015, 63-66. https://doi.org/10.1016/j.scriptamat.2015.05.021

[74] Carey, J.A., Sargent, P.M., Jones, D.R.H., Journal of Materials Science Letters, 9[5] 1990, 572-575. https://doi.org/10.1007/BF00725881

[75] Beardsley, A.L., Bishop, C.M., Kral, M.V., Metallurgical and Materials Transactions A, 50[9] 2019, 4098-4110. https://doi.org/10.1007/s11661-019-05350-6

[76] Sajjadi, S.A., Nategh, S., Materials Science and Engineering A, 307[1-2] 2001, 158-164. https://doi.org/10.1016/S0921-5093(00)01822-0

[77] Zhou, G., Li, J., Liu, C., Zhang, H., Che, X., Zhu, X., Chen, L., Philosophical Magazine Letters, 2020, in press.

[78] Sargent, P.M., Ashby, M.F., Scripta Metallurgica, 18[2] 1984, 145-150. https://doi.org/10.1016/0036-9748(84)90494-0

[79] Ashby, M.F., Acta Metallurgica, 20[7] 1972, 887-897. https://doi.org/10.1016/0001-6160(72)90082-X

[80] Kawasaki, M., De A. Mendes, A., Sordi, V.L., Ferrante, M., Langdon, T.G., Journal of Materials Science, 46[1] 2011, 155-160. https://doi.org/10.1007/s10853-010-4889-2

[81] Kawasaki, M., Lee, S., Langdon, T.G., Scripta Materialia, 61[10] 2009, 963-966. https://doi.org/10.1016/j.scriptamat.2009.08.001

[82] Mohamed, F.A., Langdon, T.G., Scripta Metallurgica, 10[8] 1976, 759-762. https://doi.org/10.1016/0036-9748(76)90358-6

[83] Ashby, M.F., Acta Metallurgica, 20[7] 1972, 887-897. https://doi.org/10.1016/0001-6160(72)90082-X

[84] Meyers, M.A., Vöhringer, O., Lubarda, V.A., Acta Materialia, 49[19] 2001, 4025-4039. https://doi.org/10.1016/S1359-6454(01)00300-7

[85] Chang, L., Zhou, C.Y., Pan, X.M., He, X.H., Materials and Design, 134, 2017, 320-330. https://doi.org/10.1016/j.matdes.2017.08.058

[86] Zhao, Q., Yang, F., Torrens, R., Bolzoni, L., Materials and Design, 169, 2019, 107682. https://doi.org/10.1016/j.matdes.2019.107682

[87] Cheng, L., Chen, Y., Xue, X., Kou, H., Li, J., Emmanuel, B., Rare Metal Materials and Engineering, 48[11] 2019, 3487-3493.

[88] Quan, G.Z., Ma, Y.Y., Zhang, Y.Q., Zhang, P., Wang, W.Y., Materials Science and Engineering A, 772, 2020, 138745. https://doi.org/10.1016/j.msea.2019.138745

[89] Tanaka, H., Yamada, T., Sato, E., Jimbo, I., Scripta Materialia, 54[1] 2006, 121-124. https://doi.org/10.1016/j.scriptamat.2005.08.045

[90] Kameyama, T., Matsunaga, T., Ueda, S., Sato, E., Kuribayashi, K., Journal of Japan Institute of Light Metals, 60[3] 2010, 111-116. https://doi.org/10.2464/jilm.60.111

[91] Quan, G.Z., Wen, H.R., Liang, J.T., Mao, A., Luo, G.C., Wang, Y.Q., Transactions of Materials and Heat Treatment, 36[4] 2015, 25-33.

[92] Quan, G.Z., Lv, W.Q., Liang, J.T., Pu, S.A., Luo, G.C., Liu, Q., Journal of Materials Processing Technology, 221, 2015, 66-79. https://doi.org/10.1016/j.jmatprotec.2015.02.002

[93] Janghorban, K., Esmaeili, S., Journal of Materials Science, 26[12] 1991, 3362-3365. https://doi.org/10.1007/BF01124686

[94] Zhou, G., Chen, L., Liu, L., Liu, H., Peng, H., Zhong, Y., Materials, 11[7] 2018, 1212. https://doi.org/10.3390/ma11071212

[95] Liu, C., Zhou, G., Wang, X., Liu, J., Li, J., Zhang, H., Chen, L., Materials, 12[21] 2019, 3520. https://doi.org/10.3390/ma12213520

[96] Ashby, M.F., Acta Metallurgica, 20[7] 1972, 887-897. https://doi.org/10.1016/0001-6160(72)90082-X

[97] Ashby, M.F., Acta Metallurgica, 20[7] 1972, 887-897. https://doi.org/10.1016/0001-6160(72)90082-X

[98] Cao, F., Materials Science and Engineering A, 643, 2015, 169-174. https://doi.org/10.1016/j.msea.2015.07.042

[99] Kawasaki, M., Langdon, T.G., Materials Science Forum, 783-786, 2014, 2647-2652. https://doi.org/10.4028/www.scientific.net/MSF.783-786.2647

[100] Kawasaki, M., Langdon, T.G., Journal of Materials Science, 48[13] 2013, 4730-4741. https://doi.org/10.1007/s10853-012-7104-9

[101] Arieli, A., Mukherjee, A.K., Materials Science and Engineering, 47[2] 1981, 113-120. https://doi.org/10.1016/0025-5416(81)90216-0

[102] Kawasaki, M., Langdon, T.G., Materials Science and Engineering A, 528[19-20] 2011, 6140-6145. https://doi.org/10.1016/j.msea.2011.04.053

[103] Kawasaki, M., Langdon, T.G., Minerals, Metals and Materials Series, Part F2, 2017, 117-131. https://doi.org/10.1007/978-3-319-51097-2_10

[104] Langdon, T.G., Mohamed, F.A., 1, 1976, 428-432.

[105] Kaddour, D., Frechinet, S., Gourgues, A.F., Brachet, J.C., Portier, L., Pineau, A., Scripta Materialia, 51[6] 2004, 515-519. https://doi.org/10.1016/j.scriptamat.2004.05.046

[106] Knorr, D.B., Notis, M.R., Journal of Nuclear Materials, 56[1] 1975, 18-24. https://doi.org/10.1016/0022-3115(75)90193-2

www.ingramcontent.com/pod-product-compliance
Lightning Source LLC
Chambersburg PA
CBHW071708210326
41597CB00017B/2382